Exercise and Eating Disorders

D1245560

Eating disorders (EDs) have become a social epidemic in the developed world. This book addresses the close links between EDs and exercise, helping us to understand why people with EDs often exercise to excessive and potentially harmful levels. This is also the first book to examine this issue from an ethical and legal perspective, identifying the rights and responsibilities of people with EDs, their families and the fitness professionals and clinicians that work with them.

The book offers an accessible account of EDs and closely examines the concept of addiction. Drawing on a wide range of medical, psychological, physiological, sociological and philosophical sources, the book examines the benefits and risks of exercise for the ED population, explores the links between EDs and other abuses of the body in the sports environment and addresses the issue of athletes with disordered eating behavior. Importantly, the book also surveys current legislation and professional codes of conduct that guide the work of fitness professionals and clinicians in this area and presents a clear and thorough set of case histories and action points to help professionals better understand, and care for, their clients with EDs.

Exercise and Eating Disorders is important reading for students of applied ethics, medical ethics and the ethics of sport, as well as for fitness professionals, psychiatrists, clinical psychologists, sports coaches and sport and exercise scientists looking to improve their understanding of this important issue.

Simona Giordano is Senior Lecturer in Bioethics at the School of Law, University of Manchester, UK. She is Programme Director of medical ethics teaching in undergraduate medical education in the School of Medicine and also teaches for the Master and Postgraduate Diploma in Healthcare Ethics and Law. Simona is a member of the UK Register of Exercise Professionals, and qualified as an exercise instructor in 1999.

Ethics and Sport
Series editors:
Mike McNamee
University of Wales, Swansea

Jim Parry
University of Leeds

Heather Reid
Morningside College

The Ethics and Sport series aims to encourage critical reflection on the practice of sport, and to stimulate professional evaluation and development. Each volume explores new work relating to philosophical ethics and the social and cultural study of ethical issues. Each is different in scope, appeal, focus and treatment but a balance is sought between local and international focus, perennial and contemporary issues, level of audience, teaching and research application, and variety of practical concerns.

Genetically Modified Athletes
Biomedical ethics, gene doping and sport
Andy Miah

Human Rights in Youth Sport
A critical review of children's rights in competitive sports
Paulo David

Genetic Technology and Sport
Ethical questions
Edited by Claudio Tamburrini and Torbjörn Tännsjö

Pain and Injury in Sport
Social and ethical analysis
Edited by Sigmund Loland, Berit Skirstad and Ivan Waddington

Ethics, Money and Sport
This sporting Mammon
Adrian Walsh and Richard Giulianotti

Ethics, Dis/Ability and Sports
Edited by Ejgil Jespersen and Mike McNamee

The Ethics of Doping and Anti-Doping
Redeeming the soul of sport?
Verner Møller

The Ethics of Sport Medicine
Edited by Claudio Tamburrini and Torbjörn Tännsjö

Bodily Democracy
Towards a philosophy of sport for all
Henning Eichberg

Ethics, Knowledge and Truth in Sports Research
An epistemology of sport
Graham McFee

Exercise and Eating Disorders
An ethical and legal analysis
Simona Giordano

Exercise and Eating Disorders

An ethical and legal analysis

Simona Giordano

Routledge
Taylor & Francis Group

LONDON AND NEW YORK

First published 2010
by Routledge
2 Park Square, Milton Park, Abingdon, Oxon, OX14 4RN

Simultaneously published in the USA and Canada
by Routledge
270 Madison Avenue, New York, NY 10016

Routledge is an imprint of the Taylor & Francis Group, an informa business

© 2010 Simona Giordano

Typeset in Goudy and Helvetica by
Swales and Willis Ltd, Exeter, Devon
Printed and bound in Great Britain by
TJ International Ltd, Padstow, Cornwall

British Library Cataloguing in Publication Data
A catalogue record for this book is available from the British Library

Library of Congress Cataloging-in-Publication Data
Giordano, Simona, Dr.
  Exercise and eating disorders : an ethical and legal analysis /
  Simona Giordano. — 1st ed.
      p. ; cm.
  1. Eating disorders.   2. Exercise.   3. Exercise addiction.   4. Eating
  disorders—Treatment—Moral and ethical aspects.   5. Athletic
  trainers—Professional ethics.   I. Title.
  [DNLM: 1. Anorexia Nervosa—psychology.   2. Athletic
  Performance—ethics.   3. Athletic Performance—psychology.
  4. Bulimia Nervosa—psychology.   5. Exercise—psychology.
  6. Liability, Legal. WM 175 G497e 2010]
  RC552.E18.G56 2010
  616.85′260642—dc22   2009041565

ISBN10: 0–415–47605–4 (hbk)
ISBN10: 0–415–47606–2 (pbk)
ISBN10: 0–203–88554–6 (ebk)

ISBN13: 978–0–415–47605–8 (hbk)
ISBN13: 978–0–415–47606–5 (pbk)
ISBN13: 978–0–203–88554–3 (ebk)

Contents

Foreword

A practical tool and a critical mirror

Walter Vandereycken

The French philosopher Michel Foucault has made the distinction between two social constructions of the modern body. The intelligible body represents the wider cultural arena of social control, whereas the useful body is the practical and direct locus of social control through which culture is converted into habitual bodily activity. Aesthetic representations of the body are translated in a set of practical rules and regulations, in particular norms of beauty and models of health. By obeying these sociocultural prescriptions, for example, through the discipline of diet and exercise, the living body is shaped into a socially adapted and 'useful' body, regulated in the interest of public health, economy, and political order.

Dietary practices and physical exercises are now aids to self-presentation. Striving for self-fulfillment, individuals depend upon validation from others. In this quest for validation, the 'self' is expected to be transparent through its physical appearance. To be a successful self in competitive social relations requires a successful body, disciplined to enhance personal value with the help of a growing sector of body-work professions (dietitians, cosmetologists, plastic surgeons) and a powerful keep-fit industry. Following the performance ethic and assisted by scientific disciplines, this successful self can be calculated using weight charts, calorie tables, and fitness schemes.

The beauty culture has become scientific, and medical sciences have promoted the rise of disciplined and useful bodies. The medical diet seeks to preserve the inner body, the 'body-machinery' (health, youth), while the consumer diet is aimed at enhancing the surface of the body, the 'face-work' (beauty, distinction). Keeping the body in good shape, then, means to make it both productive and attractive, competitive and distinctive, successful and desirable, a rational tool and a vehicle of pleasure. But at what price?

Whether corseted under tight external constraints or internally disciplined through diets and exercises, why do so many people comply with the social prescriptions of specific body sculpture up to the point of jeopardizing their own health? This book does not only offer a wealth of information and practical advice, it also faces us with challenging ethical and legal questions. And, indirectly, Simona Giordano is holding a mirror up to the reader's face. Whether fat or thin, fit or sick, instead of programming new aesthetic, scientific, or therapeutic codes, we should first try to decode the sociocultural messages of the modern body. Both fitness obsession and eating disorders force us to look in a cultural mirror and to put our society on the scales. For many of us, whether professionals in health care or fitness, this might expose the fatness of our prejudices, the fitness of our norms, and the thinness of our tolerance.

Foreword

Søren Holm

This is a book that ought to be read by all exercise and fitness professionals and by most moral philosophers and bioethicists.

But you might well ask – why should I read it? And what qualifications does Søren Holm have to make such a judgement?

Let me deal with the last question first. I have not been inside a gym nor participated in any kind of organised physical activity since I left school, so it is definitely not my first-hand experience of the modern exercise environment or industry that qualifies me to assess the merits of this book. I have, however, edited one of the most prominent ethics journals, the *Journal of Medical Ethics*, for the last five years and seen more than 1,000 papers of varying quality in that capacity. This has given me some insight into what is quality writing and argument and what is not; and this book is definitely quality!

But, you should not read this book just because I urge you to do it. I am not trying to convince you merely by an argument from authority. You should read it because it is a very good, original, well-written and, most importantly, highly practical and useful book.

For exercise professionals this book is useful because it deals with a real world problem in a practical way. How should they act when they suspect or know that one of their clients has an eating disorder? This is not a trivial problem. Eating disorders are common, persons with eating disorders are attracted to exercise as a means of burning energy and exercise can be both harmful and beneficial to those who have an eating disorder. But there is very little extant material on how to handle this problem. In this book Simona Giordano, who is both an eminent philosopher and an exercise professional herself, provides comprehensive information about eating disorders, their causes and their consequences. She also provides a straightforward and down-to-earth analysis of the ethical obligations of exercise professionals and shows how this analysis has direct implications for how they ought to act towards clients with eating disorders. This is all drawn together in a very practical table with advice to studio instructors, gym instructors, personal trainers, gym managers and fitness enthusiasts in general.

The first part of the book provides an easy-to-read overview of what we currently know about eating disorders. What are their causes and their symptoms? What are the consequences for those who suffer from eating disorders? And, are there any effective treatments? In dealing with these issues the perspective is always the perspective of the exercise professional and the treatment of competing theories of causation and treatment is generally well balanced. This part of the book gives the professional the necessary scientific and medical evidence base for making reasonable decisions. All of this knowledge is drawn together in a simple to use *identification kit* which enables exercise professionals to correctly identify clients with eating disorders.

The second part of the book then deals with how exercise professionals should deal with such clients. It provides an overview of the legal rules that apply in this context, especially the implications of the concept of *duty of care*; and also investigates the professional guidelines issued by the UK Register of Exercise Professionals, the American College of Sports Medicine, and the US Fitness Standards Council. This analysis shows that neither the legal rules nor the professional guidelines provide any significant guidance in this area. In deciding how to act the exercise professional can therefore not rely solely on rules and regulations; independent ethical judgement is thus necessary. Chapter 9 gives an exemplary and very clear exposition of the main concepts in ethics and Chapter 10 draws out the implications for exercise professionals. It is not always obvious what an exercise professional ought to do when relating to a client with eating disorders. There are ethical considerations drawing in different directions because 1) exercise can be both beneficial and harmful to people with eating disorders and 2) exercise professionals have both general ethical obligations to minimise harm and professional ethical obligations to act in the best interest of their clients. There are of course some clear cases. A client who is clearly emaciated and who is fainting during exercise should not be allowed to continue and clients should in general be screened for cardiovascular problems. But there are many more cases that are not so clear cut. In the final analysis Simona Giordano argues that there is no moral obligation for exercise professionals to assist clients with eating disorders, but that this does not mean that helping such clients is not a good thing to do. If an exercise professional is knowledgeable and competent and believes on reflection that he or she can help the client to achieve a better outcome (for instance a better relation to her or his body), then engaging with such a client is a supererogatory act, an act that is morally good but goes beyond moral duty.

This ethical analysis is significantly influenced by a series of interviews with different kinds of exercise professionals reported in Chapter 7. These interviews show a number of things: 1) that exercise professionals do reflect on their ethical, legal and professional obligations, 2) that they try to act in ways that are good and right, but 3) that their possibility to act is constrained in various ways by their own knowledge, the organisational context in which they work, the organisation's priorities, and their responsibilities towards other clients.

These interviews give those of us who are not familiar with the exercise industry an important insight into the very complex organisational environment exercise professionals have to negotiate in order to act in the way they think is the right way. In this way they enrich the ethical analysis. But the interviews are also important in another sense. Without the interviews it would be easy to forget that ethical decisions are made by real people in a context with multiple conflicting demands, and that decisions often have to be made here and now on the basis of limited information. We may later with hindsight believe that some other decision would have been better, all things considered, but hindsight is often a biased vantage point. The (alleged) fact that we would have chosen differently if we had had more time and more information does not show that the decision we did make was wrong.

Moral philosophers and bioethicists should read this book because it exemplifies a number of the virtues of good bioethics.

First, the ethical analysis is clear and comprehensive and it is furthermore not 'dumbed down' in any way. This book thus shows how in-depth ethical analysis can be communicated effectively to practitioner communities with limited or no background in ethics.

Second, the book shows how theoretical analysis can fruitfully build on and interact with empirical data from the sciences and elsewhere. To perform the ethical analysis it is necessary to have a clear picture of the 'nature' of eating disorders (their symptoms, causes, likely development, etc.) and a clear picture of the role of exercise professionals in the modern exercise industry. Understanding the role of exercise professionals not only involves a formal understanding of the context in which they work, but also an understanding of how they themselves perceive, negotiate and enact that role. The ethical analysis would just have been much more impoverished, and potentially misleading, if it had not taken account of the rich interview data presented in Chapter 7.

Third, the practical advice given to exercise professionals flows directly from the ethical analysis, but is nevertheless still eminently practical. It is no 'airy fairy' philosopher's talk; and it is not pitched at a level where it will have no application to the real world in which exercise professionals work. This is something that bioethicists wishing to have an impact on how people actually act in the real world ought to aspire to.

Fourth and finally the analysis deals with an area of practice on the fringes of the traditional core concerns of bioethics, i.e. medicine and the biotechnologies. In doing so it both shows how the borders of bioethics can successfully be pushed outwards and how standard arguments and motifs can only be applied in new contexts if they are sensitively reanalysed.

Preface

A Cimma

Ti t'adesciàe 'nsce l'èndegu du matin
ch'à luxe a l'à 'n pè 'n tera e l'àtru in mà
ti t'ammiàe a ou spègiu dà ruzà
ti mettiàe ou brùgu rèdennu'nte 'n cantùn
che se d'à cappa a sgùggia 'n cuxin-a stria
a xeùa de cuntà 'e pàgge che ghe sùn
'a cimma a l'è za pinn-a a l'è za cùxia
Cè serèn tèra scùa
carne tènia nu fàte nèigra
nu turnà dùa
Bell'oueggè strapunta de tùttu bun
prima de battezàlu 'ntou prebuggiun
cun dui aguggiuìn dritu 'n pùnta de pè
da sùrvia 'n zù fitu ti 'a punziggè
àia de lùn-a vègia de ciaèu de nègia
ch'ou cègu ou pèrde 'a tèsta l'àse ou sentè
oudù de mà misciòu de pèrsa lègia
cos'àtru fa cos'àtru dàghe a ou cè
Cè serèn tèra scùa
carne tènia nu fàte nèigra
nu turnà dùa
e 'nt'ou nùme de Maria
tùtti diài da sta pùgnatta
anène via
Poi vegnan a pigiàtela i càmè
te lascian tùttu ou fùmmu d'ou toèu mestè
tucca a ou fantìn à prima coutelà
mangè mangè nu sèi chi ve mangià
Cè serèn tèra scùa
carne tènia nu fàte nèigra
nu turnà dùa

e 'nt'ou nùme de Maria
tùtti diài da sta pùgnatta
anène via.

The poem with which I have decided to open this book is a recipe. It is written in Genoese, the language of Genoa (a port-city on the coast of west Italy). This is my translation into English:

A Cimma

You shall wake up to the indigo of the morning,
When the light has one foot on the earth, the other on the sea.
You shall look in the mirror of a pan, you will lie the broom upright in the corner
Because if the witch runs down the chimney-pot, by the time she counts every straw,
A *Cimma* is filled in and sewed.
Blue sky, soil so dark, tender meat, don't become hard
Dear meat, bed to every deed of God, with two long needles, on the tip of your toes, you will pierce it well before you bless it in herbs
Wind of an old moon, of glow and of mist,
The cleric loses his mind and the donkey its route
Smell of the sea, mixed with a light marjoram
What else to do what else to give to the sky.
Blue sky, ground so dark, tender meat, do not become dark.
Do not return hard and in Mary's name
From this pan, all the devils go away
Then the waiters come and get it
They leave the fumes of your job.
That man gives the first cut
Eat, eat, someone will eat at you.
Blue sky, dark ground, tender meat don't become dark
Do not become hard, and in Mary's name,
From this pan, all the devils go away.

A *Cimma* is the 'head' of the pig. This is one of the least appetizing parts of the pig, but in Genoa they manage to cook one of the best local dishes out of it. They say that it is the ritual that is important to the recipe, more than the ingredients. You need to start early in the morning, when the sky is still indigo, and you need a brush, next to the fireplace, to keep the witches away from the chimney. You need to stick long needles in the hard meat before you cook it, and to recite a spell, for the meat to become tender. Cooking is sorcery. And from the hardest bit, with a bit of faith, you can get the most succulent and sophisticated dish.

Food is a ritual and a religion for the people with eating disorders. Preparation of a small salad can take them a very long time. The rituals around food are familiar to those who know or care for an eating disorder sufferer. Cutting food into small pieces, placing them carefully on the plate, putting each morsel slowly inside the mouth, like the Eucharist, chewing it for ages, like a sin, choosing the right ingredients, avoiding all the 'pollutants' (e.g. fats) are necessary expedients, which make eating tolerable to the sufferer. For eating disorder sufferers *the pan is full of devils*. Only through the rituals can the sufferer cope with food. Exercise is one of the rituals that allow the sufferer to approach food. Yet, in spite of all the ceremonies, the eating disorders demons often remain trapped in the pan for a long time. Sometimes forever.

Like the recipe of *A Cimma*, eating disorders are not something of the occult, ridiculous or irrational – they are not impossible to understand. They have indeed a special 'language', which is difficult to decipher: the language of the body. And the language of the body is enriched with historical, psychological, moral and even metaphysical meanings. Yet it is a language that one can make sense of, if one finds the right keys. In *Love in the Time of Cholera*, Gabriel Garcìa Marquez wrote: "the heart has more rooms than a brothel". Those things that seem irrational, contradictory, or ambivalent often have a profound meaning, and their own rationale. Like *A Cimma*, the rituals of eating disorders have an historical background, made of values and beliefs, which largely explains why eating-disordered people get caught in the grip of their occult rituals of purification. But, as this poem also reminds us, sometimes it is actually out of the most difficult and tough situations that may come the best.

In writing this book, I have had a number of aims, or hopes, in mind: one is to assist fitness professionals and all fitness enthusiasts to deal competently and intelligently with the tragedy of eating disorders. The other is to bring to the attention of the academic world and of the public at large the ethical and legal dilemmas encountered in sports and fitness. The other is to enhance under-standing of the potential benefits and hazards of exercise for eating disorders sufferers: this is relevant to sufferers, their families, psychiatrists, psychologists and psychotherapists, and clinicians. The other is to propose modes of action with these vulnerable exercisers in the sports and fitness arena. To care for this group of fitness enthusiasts means that we should never ignore their drama and just let them be. To care for them might require asking them to stop exercising, or else, being prepared to supervise them with special skills and competencies. Finally, I hope I show that exercise and sports should never be demonized or blamed for the drama that some of us live. Exercise and physical activity can be turned, from an instrument of self-destruction, into an instrument of self-love: *from this pan, all the devils go away!*

Acknowledgments

My profound gratitude, as ever, to John Harris, who always has more faith in my ideas than I do. Thanks to Francesco Giglio, Anne-Maree Farrell, and Margaret Brazier, who have helped me greatly with the legal parts of the book. John Coggon has given extensive comments on both the legal and the philosophical parts of this book. My gratitude also to Iain Brassington, for his comments on the philosophical sections of this book, and to Adam James, who has proofread and given extensive comments on the whole manuscript. I also wish to thank Michael McNamee for his many and very useful comments on the whole book. I wish to thank Marcello Carriero for introducing me to the artworks of Vanessa Beecroft and other artists, and for the patience with which he explained their works to me. A special thank you to Walter Vandereycken, who has read the whole manuscript, kept me up-to-date with his new findings and ideas, helped me with the understanding of clinical matters, and offered many invaluable comments. I am very grateful to my colleague instructors, who have agreed to share their experiences with me for the purposes of this book. Without their cooperation, this book would not have been written. Special thanks, finally, to all my students and fitness managers with whom I have worked since I qualified as an exercise instructor in 1999. I should mention Chorlton Leisure Centre, the old McDugall Centre, the Armitage Centre, the Manchester Aquatics Centre, the Bannatyne Centre, the GL-14, and the Gymnica in Italy.

For the permission to use their material, I wish to thank Vanessa Beecroft, who has made images of her own artwork available for this book, Unilever, and the National Library of Medicine for photo permission. I also wish to thank Dr. Gregg Jantz for his permission to quote in full the story "Down the rabbit hole" and La Fondazione Fabrizio De André for permission to use "A Cimma" in the Preface.

I would like to acknowledge the stimulus and support of the iSEI Wellcome Strategic Programme in The Human Body, its Scope Limits and Future, in the preparation of this book.

Introduction

Introduction

Christy Henrich, 22, gymnast who suffered from anorexia
by Eric Pace

Christy Henrich, a former gymnast of near-Olympic caliber who had eating disorders, died on Tuesday in Research Medical Center in Kansas City, Mo., near her hometown, Independence, Mo. She was 22. The cause of death was multiple organ system failure [. . .]. Ms. Henrich had long suffered from two eating disorders, anorexia nervosa and bulimia. After narrowly failing to make the 1988 United States Olympic gymnastics team, Ms. Henrich, a hard-driving competitor who was only 4 feet 10 inches tall, placed fourth in the unequal parallel bars event in the 1989 world championships. Forced to retire at that time, she weighed 93 pounds, but then she grew frailer and thinner. Weakness forced her to retire from her sport in 1991, and in mid-1993 she was down to 60 pounds. Ms. Henrich's death came after a week in the medical center following 10 days in critical condition in a hospital in Independence, Mo. [. . .]. [H]er mother declared: "I would say 99 percent of what has happened to Christy is because of the sport".[1]

The New York Times, 28 July 1994

In 1997, Jeffrey Gantz, of the **Boston Phoenix**, reported

I must have seen Heidi Guenther perform with Boston Ballet on a number of occasions. I wish I could say I remember her. That's often the lot of ballet dancers who are corps members: even the critics scarcely notice you. It's why those dancers are so eager to graduate from the corps and become soloists and principals. Back on June 30, while spending the summer with her mother, Patti Harrington, in San Francisco, Guenther collapsed and died. The company is in shock; it's also under a microscope. The Boston Globe ran a front-page story last Thursday with the headlines "A dancer's death raises questions," "Boston Ballet had told woman to lose weight," and "Dancer, told to lose weight, dies unexpectedly." The first paragraph stated, "A 22-year-old Boston Ballet dancer who had developed an eating disorder after the ballet told her to lose weight died unexpectedly last week while home on summer break." At a packed press conference Thursday afternoon – called hastily, and obviously in response to the *Globe* story – Boston Ballet representatives stated that the company had most recently told Heidi Guenther not to lose weight, and that they had no idea she might have had an eating disorder [. . .]. The *Herald* chipped in with "Pressures, priorities and a dancer's death" and "Slender is dancers' watchword." Just back from Europe, the company's new artistic director, Anna-Marie Holmes, replied to the *Globe* that "I took as much care as I could have. But I couldn't force her to get help. There's only so much you can go into someone's personal life." Ballet in general, and Boston Ballet in particular, is likely to be under a dark cloud for some time to come. For many companies, a woman can't be too light or too thin: just a few weeks ago, former dancer Lea Thompson (*Caroline in the City*) told *TV Guide* how American Ballet Theater had advised her that at 5'5" and 96 pounds she was "too stocky" [. . .]. Questions are now being asked about what ballet companies demand of their female members, and what steps they take to protect their dancers' health. Over the past week, Boston Ballet has had to defend itself against the implication that it was responsible for Heidi Guenther's death. Is the company culpable? [. . .][2]

These are two stories of professional athletes suffering and dying from eating disorders. The pressure to succeed in professional sports might trigger the onset of eating disorders. Eating disorders, however, do not affect professional athletes alone, but all sections of the population. Around 1.6 million people are afflicted by eating disorders in the UK,[3] and around 9 million people in the US. This figure does not account for the many silent sufferers. Not only are eating disorders a worryingly widespread condition, they are also considered by some to be the most lethal of all psychiatric conditions.[4]

Exercise is one of the clinical features of eating disorders. Often the sufferer, obsessed with her/his body, and in response to an overwhelming terror of fat, exercises to burn out calories ingested with food and drinks. For fitness and sports instructors, having an eating-disordered client is now a common event (Chapter 7). This raises important ethical and legal issues that need to be addressed, and which concern in particular the relationship between fitness and sports instructors and eating-disordered exercisers.[5] These issues will be analyzed in Chapters 8 and 9. However, the issue of eating disorders touches all sports and exercise leaders, whether professionals or amateurs, as well as all of us. For example, school teachers might teach physical activity to one or more pupils with known or suspected eating disorders, or who are either too thin or obese. All leaders of recreational activities involving exercise, even if amateurs, might come across anorexic participants or enthusiasts who are evidently not healthy enough for that group activity. What should one do, for example, if one organizes a recreational event, and one of the participants has a known and severe eating disorder? Is it better to ignore the problem, in the hope that nothing will happen? Or should one address it, and, if so, how? Eating disorders and exercise also touch the public at large. You might be a gym user, and be worried that, or know that, one of your fellow users has a severe eating disorder, and s/he also seems to overdo it. Should you speak to the gym instructor? Should you get involved at all? Should you confront her/him? Which, if any, other routes are open to you?

This book will address these and other questions: What are the risks and benefits of exercise for people who have or may have eating disorders? At what levels, or can exercise be somehow beneficial to people with eating disorders? Is there a moral duty of gym users and fitness enthusiasts to help and perhaps warn fellow enthusiasts that they think may be at risk?

Eating disorders are a serious condition, and exercise, albeit potentially beneficial, if carefully designed and monitored, can be a serious hazard for sufferers (Chapter 6), especially because people with eating disorders, differently to other people with other conditions, often exercise despite their state of debilitation and regardless of professional advice to moderate exercise intensity and frequency. Eating-disordered exercisers risk cardiovascular collapse, fractures and death, depending on the duration and severity of the condition. Extreme diet, bingeing and purging, the use of compensatory methods such as self-induced vomiting and the use and abuse of diuretics and laxatives, cause severe damage to the organism, such as decreases in bone mineral density, electrolyte imbalances, reduction of the size of the heart, and so on [6] (see Chapter 2 for a full account).

In light of these potential hazards, I argue that when a participant in exercise or sports shows signs of an eating disorder, her/his problem should be addressed. It is not always possible for fitness leaders to establish the severity of eating disorders; it is therefore best to err on the side of caution and tackle the problem – rather than ignoring it, which might allow a preventable accident to happen. Later in the book, I will discuss what fitness leaders can do, when they think that one of their clients suffers eating disorders.

Fitness and sports professionals[7] face particular ethical and legal challenges. Should someone with a known eating disorder be allowed to participate in fitness and sports activities?[8] What would happen if an eating-disordered exerciser collapsed and suffered harm during an exercise session? Would the supervising instructor be responsible for that harm? What do professional codes of ethics suggest? What are the legal obligations and rights of fitness professionals towards vulnerable clients?

I shall discuss how eating disorders manifest themselves, who they mainly afflict and discuss how society at large, and fitness and sports professionals specifically, might better respond to the challenges they pose. This book unravels various ways of thinking about what fitness instructors should do in the presence of an eating-disordered exerciser. It will also discuss (Chapter 8) the legal responsibilities of fitness professionals. The ethical analysis offered in this book can be applied, however, by extension, to everyone afflicted by eating disorders – whether a sufferer or a witness to someone else's condition. It will thus also discuss what different agents (studio instructors, gym instructors, gym managers, amateur fitness leaders, and fellow exercisers) should or could do in these circumstances. Finally, I will make a recommendation that fitness bodies responsible for the education of fitness professionals design educational courses that enable advanced fitness professionals to supervise and design training programmed for people with eating disorders.

The scope and limits of ethical analysis

One of the main functions of moral philosophy[9] is to enhance an understanding of the complexities inherent in critical situations. In doing so, philosophy tries to clarify the terms and concepts and to unravel these complexities. Philosophers are often also required, or expected, to give normative answers to ethical dilemmas: thus, not only to unravel quandaries, but also to tell others what they should do. In order to do so, philosophers can analyze the different courses of actions that are open to the agents and assess their strengths and weaknesses. Sometimes there may simultaneously be various persuasive reasons for and against a certain type of conduct. In these cases, the role of a philosopher[10] is to analyze the pros and cons of the various options open to the agents, and to make suggestions that appear grounded in reasoned judgment, rather than on 'gut feelings' or 'habit' or 'assumptions'.

This is what this book proposes to do: give direct answers, wherever possible, and analyze conflicting courses of action in all other cases. This book was born out of a desire to produce a useful resource to all those involved in fitness and physical activity, including sufferers of eating disorders, their parents, exercise participants, professional sportsmen and women, instructors and gym managers, the diet-driven, elite athletes and talented sports kids.

Although I give practical advice and propose an action plan for management of the eating-disordered exerciser, I am reluctant to insist on 'norms' of behavior to be adopted towards people with eating disorders. Eating-disordered people, like all those whose difference is not fully understood, are sometimes ridiculed, and almost invariably judged for their behavior. They are often victims of paternalism, sometimes even medically treated against their will. This, as I will explain, often erodes still further the already fragile self-esteem of the sufferer. The eating-disordered exerciser does not need to be punished in any way for her or his drama. In extreme cases, I argue that a proper understanding of duty of care requires fitness professionals to request suspension of physical activity. I also argue that there is no moral obligation to supervise or coach someone with an eating disorder. However, in all cases, decisions must be based on respect for the sufferer, and understanding of the psychological and emotional dynamics that might lead the sufferer to commit self-harm. At the end of this book, I make examples of ethical decision-making towards the eating-disordered client, and propose an action plan aimed at protecting the client without being perceived as unnecessarily severe.

This book also argues that society at large has a responsibility to ensure that these vulnerable people are not lost. Exercise should not be demonized. Myths such as that exercise can be addictive or can cause eating disorders have been discredited (see Chapter 4). Exercise can be important to the resolution of eating disorders, and can be turned from an instrument of self-destruction into one of self-esteem. This transition requires an open attitude towards eating disorders, and an open and informed liaison between sports people, including professionals and health professionals. But it also requires that fitness-regulating bodies, which are responsible for the continued education of fitness professionals, such as the Register of Exercise Professionals (REPs) in the UK or the American College of Sports Medicine (ACSM), prepare fitness professionals to competently and satisfactorily assist these clients, and that they train fitness professionals to be capable of monitoring exercise and exertion specifically for the eating-disordered exerciser.

Eating-disordered exercisers: How to identify them

It is often difficult to identify the person with an eating disorder. Many eating disorder sufferers have a normal weight; some might even be slightly overweight; many either deny that there is anything wrong with them, or admit it and claim that the problem is under control. Even at normal weight, someone with eating disorders can be at serious risk of physical damage (see earlier in this chapter and also Chapters 1, 2 and 6). Of course, nobody expects a fitness amateur to spot all participants at risk: if my brother, who is not a fitness professional, organizes a five-a-side in his back garden, and one of his friends has a non-evident eating disorder, which she has never mentioned, of course my brother cannot be expected to identify her. However, fitness and sports instructors should, in some cases, be able to identify eating disorders. This is also a part of their duty of care (see Chapter 8). How can fitness professionals identify those at risk? This book provides *an identification kit*, specifically designed for professionals involved in the fitness and sports industry, but usable to everyone concerned with, or touched by, eating disorders. It gives a clear account of signs and symptoms of eating disorders, and thus helps anyone in the fitness and sports arena to recognize the eating-disordered exerciser.

Legal issues

The presence of the eating-disordered exerciser poses important legal dilemmas for fitness and sports professionals. Would fitness professionals be liable for any injury incurred by the eating-disordered exerciser? Should fitness professionals request medical certifications? And what is the legal role of medical certifications in protecting the instructor against liability?

Whereas fitness instructors have a duty of care towards their clients, how this duty should be understood is not immediately clear. Chapter 8 explains the legal concepts that can assist professionals involved in making informed decisions relating to eating-disordered exercisers. It clarifies in which cases fitness professionals could be held liable for breach of duty of care, or liable in negligence in case of harm to the client. It also reviews the most important international codes of ethical practice and standards of care, which will assist instructors in understanding their legal rights and responsibilities.

Although the law gives important indications as to how instructors should deal with people at risk, it does not resolve the major dilemmas raised by the presence of eating-disordered exercisers, and leaves it to the discretion of the fitness professionals as to how to handle situations of risk. This means that fitness and sports instructors are charged with the resolution of profound practical dilemmas, which are primarily ethical (rather than legal) in nature.

Ethical dilemmas

Fitness professionals are generally not medically qualified, and normally do not have the expertise required to recognize, minimize or reverse the secondary symptomatology of anorexia and bulimia, yet eating disorders sufferers populate the gym and the studio. This is a state of affairs that cannot be changed unless the fitness industry implements draconian measures. If barring everyone who is suspected of having an eating disorder is not a desirable option, then one has to accept that there are a number of eating disorders sufferers among the clients or supervisees of many fitness and sports instructors: some are more severely affected, others will have subclinical symptoms. Some will talk about their condition, others will deny it. Under these circumstances, it is desirable that the professionals involved become able to identify eating-disordered exercisers, wherever this is possible, develop skills and plans to approach these exercisers, and are enabled to obtain qualifications to coach them properly. Turning a blind eye to a difficult situation is not a good solution.

It is well known to those involved in the management of eating disorders that caring for an anorexic is demanding, because something within the anorexic pulls her in the direction of self-destruction, and she will strenuously defend the disordered habits (see Chapter 1). The classic moral conflict between respect for self-determination and paternalism, which afflicts the care and management of eating disorders sufferers in health care, expands itself to the fitness and sports arena.

US fitness and sports bodies, such as the ACSM, refer to fitness professionals as *health professionals*. As such, they have an imperative to promote people's health. However, like all other health professionals, they also have an obligation to respect the self-determination of their clients. A new category of health professional enters in its full legitimacy in the classic moral debate on whether, and if so why and how, people's freedom can (or should) be legitimately restricted *for their own good*. People are normally free to harm themselves, be it through smoking, drug use and abuse, alcohol consumption and, of course, overeating and exercising *to defective levels*. If a sedentary obese person cannot be forced to diet and take exercise, based on the principle of respect for self-determination, why should an eating disorders sufferer be forced to eat and to avoid exercise? Why should some people (many people – consider that there are roughly 92 million obese people in the US alone[11]) be free to ruin their life by eating to excess and not exercising enough, and others should be forced to eat and not to exercise?[12] On the other hand, should fitness instructors comply with sets of goals that do not fit into any proper understanding of their function?

The experience of fitness professionals

In the course of background research, I asked a number of fitness professionals to tell me their stories of being involved with exercisers with eating disorders (Chapter 7). Each of these fitness professionals was able to narrate stories concerning exercisers who appeared anorexic or who were known to be anorexic or bulimic. This gives an idea of the pervasiveness of the phenomenon in the fitness industry. Each fitness professional I have interviewed has thought about the dilemmas raised by eating-disordered exercisers in a different way. Some among them believed that eating-disordered exercisers should be allowed to exercise, some that they should be barred. The arguments produced are in all cases forceful and they all seemed, at first glance, persuasive. Yet, with further scrutiny it becomes clear that there are better approaches than others.

Chapter 9 analyzes these various ways of thinking about how one should behave towards the eating-disordered exerciser, and compares them. I will give *my solution* to the dilemma of whether (and if so how) we should allow people with eating disorders to take supervised exercise. I recognize, however, that there are strengths in each of the avenues of moral thinking proposed by the instructors interviewed. Part of the reason why definite generalizable answers or clear norms applicable across the board cannot be provided is that we lack data about the real risks that eating-disordered exercisers run. A study of the harm suffered globally by eating-disordered exercisers as a consequence of fitness or sports activities is not yet available. Whether such a study could be carried out is an open question. There are of course methodological and ethical difficulties involved in a study of this kind, for example, recruitment of a sufficiently large sample of sufferers who would be willing to participate in a long-term study. More importantly, there would be ethical issues relating to the exposure of the research subjects to preventable and possibly major risks.

In spite of a lack of systematic data on the effective risks, theoretical arguments for and against exercising while running 'some important' risks for health can be produced. The theoretical analysis assists us in determining whether, in absence of precise clinical data on risks, physical activity can ethically be allowed, and in what circumstances.

In Chapter 9 I argue that there are important moral reasons to accept an eating-disordered participant in exercise: these are ethical reasons grounded in beneficence, and relating to the minimization of short- and long-term harm and maximization of long-term good. Fitness professionals have good ethical reasons to accept eating-disordered exercisers. However, I argue that, in spite of these good ethical reasons, fitness professionals do not have an obligation to accept these participants under their supervision or in their center. I therefore argue that accepting an eating-disordered exerciser can be a *supererogatory* action – an action that is not morally mandatory, but morally generous. In the most severe cases, duty of care requires a request to suspend activities.

These suggestions might contradict the feelings of many fitness professionals. The case histories reported in Chapter 7 show how many fitness instructors are willing to supervise disordered clients. Yet I contend that the physical vulnerability of eating-disordered exercisers is a good reason to take radical initiatives, such as requesting suspension of activities, to protect the client's health and safety. Moreover, the average fitness instructor is typically not qualified to deal with these situations of extreme vulnerability. By accepting an eating-disordered exerciser, therefore, the professional accepts to do a job for which they are not necessarily equipped. This is why, although there might be good ethical reasons to allow an eating-disordered exerciser to take physical activity, there is not a moral obligation to do so.

In this book, I purposely refer to eating disorders as 'a condition', because it is contentious whether eating disorders are 'an illness'. I suggest that they are a condition, a state of being, rather than an illness. However, eating-disordered behavior results in a number of medical dangers. Even at basic levels of education, fitness and sports professionals should be given information about eating disorders, and about how to handle situations that might fall beyond their professional competencies. It is also important that the public at large, and the eating disorders sufferers themselves, know that many fitness professionals might not be aware of the health perils associated with abnormal eating, and might not be qualified to design and supervise exercise specifically for them.

Short summary of chapters

Chapters 1 and 2 explore the psychological, physiological and sociological dimensions of eating disorders. Incidence, prevalence and mortality rates are reported. The main aim of this part of the book is to provide a comprehensive and accessible account of eating disorders. This should enable all those who may be interested in eating disorders, including academics, people with eating disorders, their families, fitness professionals, teachers, sports coaches, sports people, clinicians, psychiatrists and psychotherapists, to have a better understanding of this condition.

Chapters 3, 4 and 5 analyze the possible causes of eating disorders. Chapter 3 explores the biology and genetics of eating disorders and assesses whether eating disorders are caused by a biological dysfunction. Chapter 4 analyzes the relationship between eating disorders, exercise and addiction. Chapter 5 explores whether eating disorders might be the result of social pressure, spread by the media, to be thin.

These three chapters explore whether eating disorders are the result of some chemical imbalance, or of some defective gene, or of an addiction, or of social pressure, or are rather a chosen pattern of behavior. This has ethical implications: if eating disorders are the result of genetic disorders, or other physical or social forces, then in principle paternalism may be justified. It would be unethical (see Chapter 9) to let someone kill her/himself while acting not out of a choice but compulsion, an addiction or some uncontrollable influence. But if there is no evidence that the eating-disordered exerciser is compelled to abuse her/his body, not only do we need some other way of understanding this condition, but also, from an ethical point of view, paternalistic intervention is much more difficult to justify.

Chapter 6 analyzes the role of exercise in eating disorders. It explains why people with eating disorders opt for certain forms of exercise, by providing a clear account of how different forms of exercise can help to burn fats and calories, and increase basal metabolic rates. It also highlights the dangers and potential benefits of exercise for eating-disordered people.

Chapter 7 gives a voice to fitness professionals. A number of case histories are presented and analyzed. I have talked to three categories of fitness professionals: personal trainers and gym instructors; studio instructors; and managers. Personal trainers and gym instructors tend to work on a one-to-one basis. Studio instructors are qualified to work with groups, and typically with averagely healthy adults. Gym managers, or managers of sports and fitness centers more generally, are not necessarily qualified in fitness and sports. However, they have been included, because they are also involved directly in the protection of the safety of the premise and the health and safety of all visitors. They also select clientele: they issue and renew memberships and might be responsible for barring clients. We will see what the experiences of these three groups are. Their narratives provide important food for thought, and the ethical analysis will take shape on the basis of these case histories.

Chapter 8, as previously described, discusses the legal rights and responsibilities of both gym managers/owners and fitness instructors.

Chapter 9 examines the ethical principles of *respect for autonomy* and *beneficence* as applied to eating-disordered exercisers. It distinguishes between *beneficence as maximization of short-term health*, as *maximization of long-term health* and *beneficence as minimization of harm*.

The final chapter provides a series of practical suggestions:

* in what circumstances it is right to bar an exerciser from physical activity;
* how to deal with our own feelings when asking an exerciser to stop exercising;
* how exercise can help a person with eating disorders;
* advice for fitness professionals;
* advice for a person with eating disorders; and
* advice for all those involved with eating disorders.

The conclusions gesture towards the extension of these arguments for all fitness enthusiasts, exercisers and sports persons who might exercise to the detriment of their health, and those with other health problems who choose to exercise at risk to themselves. The final part of this book also outlines some of the methods of analysis employed in this book for the field of applied ethics in general. In particular, it brings to the surface how philosophical and ethical analysis can be employed to examine the experience of professionals to yield practical policy recommendations.

1 Eating disorders
Symptoms and facts

Anorexia and bulimia nervosa: The terms

This chapter offers a comprehensive and accessible account of eating disorders. The term 'eating disorders' refers to a broad group of disorders, which includes anorexia nervosa, bulimia nervosa, obesity, binge eating, extreme dieting, fasting–bingeing cycles and other forms of subclinical anorexia and bulimia nervosa. By 'eating disorders' I will, however, mainly intend anorexia nervosa and bulimia nervosa (I will often use 'anorexia' and 'bulimia' to refer to 'anorexia nervosa' and 'bulimia nervosa'), which are those eating orders mainly considered in the clinical literature.[1] I shall, however, also discuss briefly what these other eating disorders, such as binge eating, are. The points raised on anorexia and bulimia nervosa can help with a more general understanding of other (sometimes less extreme) forms of disordered eating.

Anorexia nervosa

The term 'anorexia nervosa' was coined by William Gull in 1873. In 1888 this illustration appeared in his article 'Anorexia Nervosa', published in the *Lancet*.[2] However, this phenomenon was noticed much earlier by Richard Morton, in London. In the *Treatise of Consumptions* (1694), he talked of:

> A Nervous Atrophy, or Consumption [which is] a wafting of the body without any remarkable Fever, Cough or shortness of breath [. . .] In the beginning of this Disease the state of the Body appears oedematous and bloated, and as it were stuffed with dispirited Chyle; the face is pale and squalid, the Stomach loathes every thing but Liquids, the strength of the Patient declines at that rate [. . .] The immediate cause of this Distemper I apprehend to be in the System of the Nerves [. . .] The Causes which dispose the Patient to this Disease, I have for the most part observed to be violent Passions of the Mind, the intemperate drinking of Spirituous Liquors, and in unwholsom Air, by which it is no wonder if the Tone of the Nerves, and the Temper of the Spirits are deftroy'd. This Distemper as most other Nervous Diseases is Chronical, but very hard to be cured, unless a Physician be called at the beginning of it.[3]

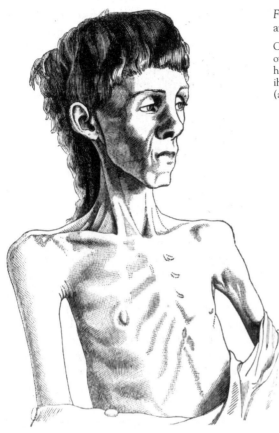

Figure 1.1 Image of an anorexic from 1888.

Courtesy of the National Library of Medicine. Available at: http://www.ihm.nlm.nih.gov/ihm/images/A/12/524.jpg (accessed 23 November 2009).

Anorexia is a word of Greek origins, and literally means lack of appetite. Contrary to what the name originally denoted, anorexia nervosa is not primarily characterized by lack of appetite. The sufferer fights against hunger. Mara Selvini Palazzoli explains that "anorexia is not *primarily* a lack or a perversion of appetite, but an impulse to be thin, which is *wanted and completely accepted by the sufferer*".[4] The anorexia sufferer engages much of her energies in this constant fight against food, hunger and absorption of calories. The more she fights against food, the more food becomes important to her. Some anorexics are also vomiters. This condition is sometimes called 'bulimic-anorexia'. People with anorexia sometimes binge, and like other people in a condition of malnutrition (prison or natural starvation), they constantly think of food.[5] Food becomes an obsession. Their entire life rotates around food, food avoidance, choice of 'safe' foods, and ways in which ingested calories can be expelled.

People with anorexia and bulimia nervosa have a terror of fat and weight. Like all fears, this fear is dominating, and the person cannot find reassurance in others' attempts to comfort. The experience of the sufferer, especially of the bulimic-anorexic, or of the bulimic, is to be trapped in a cycle that s/he cannot break. However, anorexia is not a condition that just 'happens' to the person. The urge to be thin and light is, in an important sense, deliberate. The sufferer willingly fights against hunger, and deliberately defends her eating and exercise habits. The sufferer thus plays an important part in the initiation and perpetration of eating disorders. This does not mean that sufferers should be 'blamed' for their condition – it rather means that, as they have the strength to fight against hunger, one of the most basic and compelling instincts, they also have the strength to adopt healthier lifestyles and to recover.

Bulimia nervosa

Bulimia etymologically means 'ox-hunger'. Again, the term is inadequate to capture the drama that lies behind bulimia. The person with bulimia does not binge herself or himself because s/he is very hungry. S/he binges because of an urge, experienced as being out of control, to consume as much food as possible. The person will feel terrible about bingeing. S/he will feel dirty, guilty and above all profoundly ashamed. S/he will only be released from this anguish by getting rid of ingested food. The quickest way to achieve this is self-induced vomiting. Other practices aimed at getting rid of calories include use and abuse of laxatives and diuretics, self-induced vomiting, use of appetite suppressants and exercise.[6]

As neither starvation nor binge eating are caused by dysfunctions of appetite, the obvious questions then are: Why do sufferers starve? Why do they binge?

Eating disorders sufferers, whether anorexics or bulimics, are always overwhelmed with fears relating to body shape and weight, and invariably have a terror of fat. These fears – like any other fear – are experienced as being out of willful control. Consequently abnormal eating patterns are also experienced as being out of willful control. For this reason, as I will argue in later chapters, eating disorders have often been compared to addictions and other disorders of control of impulses (previously called obsessive-compulsive disorders). The person is obsessed with the fear of fat and with the thought of food and body weight. I will discuss various hypotheses on the causes of eating disorders in Chapters 3 to 5.

Binge eating disorder

Binge eating disorder refers to the condition of those who engage in food orgies of a bulimic type, but, differently from anorexia and bulimia, will not adopt compensatory behavior. At least some of the psychological dynamics involved in binge eating seem similar to those involved in bulimia. Eating is experienced as shameful and generates strong and sometimes unbearable feelings of guilt. However, people suffering from binge eating may tend to be overweight, whereas anorexics, bulimics or bulimic-anorexics might be underweight or have normal weight.[7] Therefore, the physiological characteristics of binge eating, including the secondary symptomatology that results from binge eating, are generally different to those typical of anorexia and bulimia.

Obesity

Obesity is defined as "a condition in which the natural energy reserve, stored in the fatty tissue of humans and other mammals, is increased to a point where it is associated with certain health conditions or increased mortality".[8] Obesity is defined and measured in various ways (see Chapter 6, 'Globesity').

Obesity may also be included among eating disorders in that it results from 'disordered' and self-destructive eating. Interestingly, whereas other eating disorders are classified as *psychopathologies*, obesity is not generally regarded as such, and is not included among psychiatric diseases. But what is the demarcation line between binge eating and obesity? What is it that makes binge eating a psychological or psychiatric condition, while obesity is not? Hilde Bruch,[9] one of the first and still one of the most important scientists to have studied eating disorders extensively, in effect considered obesity an eating disorder. She showed that, in most cases, the causes of being overweight are psychological, not endocrinological or metabolic. This raises the issue of why some eating disorders are classified as mental illnesses, whereas other eating disorders (for instance, over-eating generally associated with obesity) are not.

This book is primarily concerned with anorexia and bulimia nervosa, both because these are most typically regarded as the 'proper eating disorders',[10] and also because it is anorexia and bulimia that generate the most acute problems in the fitness and sports environment (see Chapter 6).

Anorexia athletica

The term 'anorexia athletica' was first used in the 1980s by Smith[11] and Pugliese.[12] It

> describes the continuum of subclinical eating behaviors of athletes who fail to meet the criteria for a true eating disorder, but who exhibit at least one unhealthy method of weight control; this includes [. . .] vomiting [. . .], and use of diet pills, laxatives, or diuretics [. . .].[13]

Anorexia athletica is not a diagnostic category, and is not included in official diagnostic manuals.

In Chapter 6 I will return to anorexia athletica, to examine how the attempt to achieve excellence in sports and dance can trigger eating disorders. However, eating disorders and disordered exercise are not just problems for elite athletes or specific categories of professionals. They affect the population at large: young females mainly, but also older women and men and children.[14] It is on the general population that this book shall focus. I shall discuss how exercise might represent a further threat for those already at risk, due to eating disorders, and the ethical dilemmas that exercise, connected with eating disorders, can pose to fitness professionals who work in the community.

How eating disorders manifest themselves

Sarah's story[15]

> I am 23 years old, and I am still battling against anorexia. I grew up in a 'normal' family. My parents are still together, and my older sister is married. There is nothing apparently wrong with my family . . . .Yet, for me it has always been a painful situation. My parents have all the 'normal' conflicts that most couples seem to have. But it often feels . . . I am out of place. I think my anorexia started . . . at dinner time. We all had to sit around the table every night for dinner. An apparently nice family habit. But nobody could ever talk, because the news was on or dad was too tired. Nobody ever asked how the others were, or how the day had gone. There was silence, and under that, unspoken tension and unhappiness. Dinner times were pretty much the only times in the day where we'd all get together. And it was painful. My parents were very resentful towards each other, and anyone talking risked causing an outburst of anger, of shouting and insulting each other. So, we'd all remain silent.

In many cases, eating disorders sufferers complain about family situations in which love and emotional security are lacking. The relationship with parents seems an important determinant of eating disorders (this will be discussed further later in this chapter).[16]

> My unease was largely unnoticed. I started cooking 'healthy' meals for myself, or making excuses not to sit at the table with my parents. I think, at the back of my mind, I wanted to disconnect myself from what was happening around me. I bought myself a few books on nutrition and I was interested in understanding facts about food. I started experimenting with healthy recipes. Healthy cooking was good, and initially it made me feel better and more worthwhile. Eventually, however, I found it really hard to eat what my mother was cooking, that is the 'normal' food that I had always had.

Eating disorders generally begin with a diet aimed at improving eating habits or just losing a few pounds in weight.

> When I went to college, I continued with my healthy eating habits. I was studying uncompromisingly hard. It wasn't about losing weight. It was more about having a routine and ensuring I was good, healthy and successful. Everything in my day had to follow a routine. Alarm clock, breakfast, reading time, snacks, writing time, music time, jogging time . . . . My academic results were excellent, and my parents were very proud of me. I was proud of myself,

because they seemed happy with me and my results. I also started doing some voluntary work in the local community, and that was enjoyable, at least initially. Yet, I wasn't happy . . . .

Sarah, like many of those affected by eating disorders, has a middle-class upbringing, is able to go to college and is an excellent student. She shows commitment, is successful and has a strong sense of moral duty towards others. Understanding people's moral values and beliefs is essential to capture the meaning of eating disorders and over-exercise: control of the body and physical purity become the emblems of strength of will, of spiritual power over the potentially dangerous impulses of the body. A certain conception of the human person and of moral goodness is essential to understanding eating disorders and over-exercise (see Chapter 5).

The problems appeared during my second year at college. I was losing weight. Controlling every aspect of my life was good, but there is no evident limit to it. If you decide you can control your appetite by allowing yourself 60 grams of carbohydrates and a bowl of vegetables for dinner, you can increase that control by reducing your portion by 10 grams . . . see how it goes . . . . Jogging can go from 20 to 30 minutes per day . . . from once in the morning to twice a day . . . and the three apples you have for lunch can at times become two . . . see how it feels . . . .

Sometimes my friends raised some concerns about my weight. But I had no trust in anyone, and I had no trust in my physical sensations. If someone ever said to me 'you look good today' I was petrified: I thought, maybe I have put on weight, and was miserable until I could check my weight on the scale . . . I was tired and always cold. But tiredness, to me, was a sign of my weakness (not a sign of hunger). It meant I had to work out a little bit harder . . . .

The more unhappy I was with how I looked, the more I persevered in my diet and in my routines. At that point, my body and my achievements became all there was to my life. There was no room for anything else. I think at that time, I was fully anorexic.

Typically, the person with eating disorders becomes more and more absorbed with the self, and self-mastery pervades every activity and moment of the day. The warnings of others are rejected. The sufferer tends to misinterpret friends' and family's expressions of concern.[17] People with eating disorders do not trust their own bodies, and just as they do not trust themselves they do not trust most of the people around them. They only trust their rigid routines, and the more they become miserable the more they delegate to their rigid routines the resolution of their drama: in doing so, they trap themselves in what Hilde Bruch has effectively called 'The Golden Cage'.[18]

Jude's story

My name is Jude and I've had eating disorders for 13 years. My road to recovery has been about eight years long. I learned to diet from my mother. She was always on a diet. My parents were very controlling. They wanted to know everything about me, and I was seldom allowed to make any decision by myself. All my friends, activities, likes and dislikes had to be known to them and approved by them. When my parents split up I was about 14 years old. I started scheduling all my activities, friends, school work, meals. I think it was a way of claiming back the control over my life that I never had. I became obsessed with diet and exercise. Swimming and running became part of my daily routine. My parents initially ignored my new self. They put it down to the break up. Soon they realized they couldn't control it: they became frustrated, angry and worried. And I felt, probably for the first time, I had some power over them. I was very proud of my dieting, and I was proud of its effects.

Jude presents another typical story of eating disorders. In this case, the problem begins in puberty, and has a long natural cycle (which is generally between one and eight years).[19] The sufferer somehow relates the onset of eating disorders to the parents' behavior: in this case, her mother's obsession with diet and, perhaps more significantly, both her parents' attempts to direct her life, which she experiences as suffocating. In her own re-articulation of her story, Jude suggests that, as a reaction to her parents' control, she takes control over the one area of her life which nobody can direct: her feeding and her body. Indeed, Hilde Bruch, in her pioneering study of eating disorders, argued that the fight against hunger expresses emotional needs that the anorexic is unable to express in other areas of her life.[20]

Interestingly, Jude says her parents' reactions to her overly strict diet and exercise routines make her feel good about herself: they *empower* her. She thus gets *double control*, over the self, and over her parents who are now besides themselves with worry over their starving and self-harming child.

Another important aspect of this narration is the concept of *deserving* to fast. As we also saw in the previous case history, the eating disorders sufferer has absorbed a moral ideal according to which mastery of the body is somehow *good*, morally good, as it is an indication of a capacity to control low-grade and degrading impulses coming from the body (see Chapter 5). Fasting thus is not just a renouncement, but a victory. This is how being empty can be experienced as a 'fulfilling' experience, and the worthless (in the dynamic of "I am not worth feeding", "it's not worth dying for thinness") becomes worthwhile ("I am only worthwhile if I don't eat", "fasting is worthwhile, even if I will die as a result"). This is the paradox of eating disorders, a condition made of opposites that feed each other (control/lack of control; sense of worthlessness/empowerment; empty/full; wanting to change/not wanting to change; seeking help/refusing help).

When my weight became dangerously low, I was hospitalized. I gained weight quickly, but, emotionally, it was draining. Eating and putting on weight did not help me to cope better with my feelings. I still wanted not to eat, and wanted it more than anything . . . .

Jude's account tells us something important about the efficacy of coercive treatment. Forced treatment can either be a forced diet ("if you don't eat we shall keep you in hospital") òr artificial nutrition (most typically, naso-gastric feeding). Forced treatment can save someone's life, but it often does not help the patient to recover.[21] Many do not fully get over eating disorders, and continue to live with the fear of being inadequate. To cope with this feeling, many negotiate with themselves an acceptable (and generally low) weight threshold, over which they are not allowed to go.[22]

The biggest problem, as I saw it, started the day I couldn't hold it any more. One night I was just very, very hungry, and I ate all the meal that my mother prepared. She was happy to see me eat. But I felt terrible. For the first time in ages I felt 'full' and it was terrifying. I just couldn't hold it inside. I went to the toilet, and as quietly as I possibly could, I stuck a toothbrush down my throat and made myself sick. From then onwards, throwing up became my ally. Sometimes I couldn't wait till mom was out, I thought about it all day, and when she'd gone, I'd eat whatever I wanted, whatever I found, and then throw it up . . . . I cannot begin to tell how disgusted and ashamed I was at myself for this. I would have rather died than someone finding out what I was doing.

After a period of intense restrictive diet many eating disorders sufferers break through their regime. This, I have argued elsewhere,[23] can be seen as a rebellion of the 'healthy' side to the unhealthy 'controlling' side of the self. The main problem is not the breakthrough, but the rigid controlling part of the self. However, the eating disorders sufferer does not see it this way: he or she feels profoundly ashamed and guilty for eating the forbidden food. Many try to purge themselves, at least in part, in order to release the anxiety caused by the binge. Often, this is the way in which bulimia kicks in, and this is how many learn to cope with excessive unsustainable diet, with the uncontainable impulse to eat, and with the terror of putting on fat.

These stories also highlight another important aspect of eating disorders: the experiences of eating disorders sufferers are similar in many ways. Sufferers invariably have deep anguish relating to everything they eat and to their body; they come from a family that, even if in appearance is 'normal', they say has not been able to convey the 'right' type of love to them. Finally, these stories show how, unfortunately, eating disorders are almost always difficult to resolve.

This similarity in experiences and clinical picture is unusual in clinical psychology and psychiatry. Two individuals with the same diagnosis of, for example, paranoid schizophrenia, will present very different clinical pictures. Eating disorders are instead a well-defined syndrome.[24]

Box 1.1 Anorexia nervosa, from the Diagnostic and Statistical Manual of Mental Disorders (DSM-IV, Text Revision). Another official classification can be found in the ICD-10.

307.1 Anorexia nervosa. Diagnostic criteria

Early signs may include withdrawal from family and friends, increased sensitivity to criticism, sudden increased interest in physical activity, anxiety or depressive symptoms.

A. Refusal to maintain body weight at or above a minimally normal weight for age and height (e.g. weight loss leading to maintenance of body weight less than 85 percent of that expected; or failure to make expected weight gain during period of growth, leading to body weight less than 85 percent of that expected).
B. Intense fear of gaining weight or becoming fat, even though underweight.
C. Disturbance in the way in which one's body weight or shape is experienced, undue influence of body weight or shape on self-evaluation, or denial of the seriousness of the current low body weight.
D. In postmenarcheal females, amenorrhea, i.e. the absence of at least three consecutive menstrual cycles. (A woman is considered to have amenorrhea if her periods occur only following hormone, e.g. estrogen, administration.)

307.51 Bulimia nervosa. Diagnostic criteria

A. Recurrent episodes of binge eating. An episode of binge eating is characterized by both of the following:
   • eating, in a discrete period of time (e.g. within any 2-hour period), an amount of food that is definitely larger than most people would eat during a similar period of time and under similar circumstances
   • a sense of lack of control over eating during the episode (e.g. a feeling that one cannot stop eating or control what or how much one is eating)
B. Recurrent inappropriate compensatory behavior in order to prevent weight gain, such as self-induced vomiting; misuse of laxatives, diuretics, enemas, or other medications; fasting; or excessive exercise.
C. The binge eating and inappropriate compensatory behaviors both occur, on average, at least twice a week for three months.
D. Self-evaluation is unduly influenced by body shape and weight.
E. The disturbance does not occur exclusively during episodes of anorexia nervosa.

The DSM-IV distinguishes further between restricting and purging type anorexia and bulimia. Restricting anorexia is characterized mainly by fasting, whereas in purging type anorexia, the sufferer also uses other compensatory behaviors, such as self-induced vomiting or the misuse of laxatives, diuretics, or enemas. Compensatory behaviors are also called 'cathartic practices'. Bulimia is also classified further in as either purging type, in which the person has regularly engaged in self-induced vomiting or the misuse of laxatives, diuretics, or enemas; and/or non-purging type, in which the person has used other inappropriate compensatory behaviors, such as fasting or excessive exercise, but has not regularly engaged in self-induced vomiting or the misuse of laxatives, diuretics, or enemas.

Invariably, eating disorders involve feeling terrified of fat and body weight.[25] According to research, about 50 percent of women with eating disorders exercise to excess.[26] The eating disorders sufferer might avoid social situations that involve meals, including meals with the family. Family dynamics often change dramatically because of the presence of eating disorders. Family members become rightly worried for the health and life of the sufferer, but might also feel deceived and betrayed by the sufferer who does not share his or her problems with them and, instead, closes access to her inner world. She consumes her body while declaring that 'nothing is going wrong' in her life.

Denial is an inherent feature of eating disorders. Many have suggested that denial of emaciation might result from a disorder in perception of body image. Research points in different directions.[27] To assess the estimation of body size, various techniques can be used.[28] These techniques give discordant results,[29] and some report no disturbance in estimation.[30] Vandereycken suggests that this discordance might depend on the fact that these types of studies rely on patients' self-reports, and gathering true statements from eating disorder sufferers is nearly impossible.[31]

In the most severe cases, the emaciation is so extreme that it threatens the sufferer's life. In these cases, sufferers might be hospitalized (either voluntarily or involuntarily, for example, in England and Wales under the Mental Health Act 2007[32]), and subjected to re-feeding programs. These could be implemented either through persuasion or coercive measures, such as artificial feeding. Artificial feeding can be administered through a naso-enteric tube, intravenous feeding or a gastrostomy feeding tube.[33]

Bulimia often arises after a period of restrictive diet. Bulimia is more difficult to detect than anorexia. Whereas the anorexic is typically emaciated, the bulimic might have normal weight and even be slightly overweight. This might cause particular problems in the fitness environment: whereas detecting someone with anorexia can be relatively easy, identifying someone with bulimia can be much more difficult. The fitness professional needs to look at various signs of eating disorders: not just thinness, but body weight variations, evidence of excessive preoccupation with body weight and fat, or patterns of exercise. These and other indicators will be discussed in the course of this book.

The relationship between anorexia and bulimia nervosa is debated in clinical psychology and psychiatry.[34] Some argue that there are important differences between the two syndromes, and that they should be considered as separate conditions. However, the vast majority of experts consider anorexia and bulimia nervosa as two related syndromes. They often appear together, and they are the two sides of the same coin: a relentless preoccupation with thinness.[35]

Eating disorders as a 'social epidemic'

Richard Gordon talked about eating disorders as a social epidemic, a term that captures the spread of these devastating conditions.[36] In the 1970s, eating disorders were a rarity.[37] In 2005, the Eating Disorders Association, one of the UK's largest non-profit organizations, established to help people with eating disorders, estimated that 1.15 million people suffered from eating disorders in the UK.[38] Childline support,[39] another organization active in the UK, helps up to 1,300 children with eating disorders every year.

To these numbers, we should add the population of 'silent sufferers' who will never seek professional help and who will consequently not appear in clinical estimates; and people who have a difficult relationship with food and weight but fail to meet the diagnostic criteria (sometimes, for example in the DSM-IV, referred to as Eating Disorders Not-Otherwise-Specified, or EDNOS).

Not only does the disorder affect an alarming number of people, it is also considered as the most lethal of the psychiatric disorders. Mortality associated with the disorder is reported as up to 20 percent.[40] Many deaths result from suicide.[41] This gives an idea of the unbearable suffering that people with eating disorders experience.

Who develops eating disorders?

According to the DSM-IV TR, anorexia nervosa affects about 1 percent of young women, and bulimia affects about 17 percent of college-aged women (of these, 30 to 80 percent have had a history of anorexia).[42] According to the ICD-10, the condition "occurs most commonly in adolescent girls and young women, but adolescent boys and young men may also be affected, as may children approaching puberty".[43]

Originally, eating disorders manifested nearly exclusively in middle-class adolescent women, most often appearing either in schools or colleges or in particular environments, such as among dancers or people involved in the fashion industry,[44] where body shape and weight are particularly important. There is also a link between urbanization and increased prevalence[45] of eating disorders.[46]

That eating disorders were a 'female' condition became immediately evident to those who first observed the phenomenon. For example, Gull wrote: "The subjects of this affliction are mostly of the female sex and chiefly between the ages of 16 and 23."[47]

Over time the disorder has increasingly affected older people and males.[48] Males are now thought to represent around 8 percent of the anorexic sufferers, 15 percent of the bulimic sufferers and 20 percent of binge eating disorder sufferers.[49]

Anorexia generally appears most often in adolescence or early adulthood, and bulimia tends to appear later. This might be because bulimia sometimes begins after a period of restrictive anorexia. It also seems that males on average develop eating disorders slightly later in life than women.[50]

Before we look at the physiological effects of abnormal eating, I will briefly sketch the features of the families and societies in which eating disorders arise. There seem to be traits which these families and societies share. Awareness of these traits might help sufferers and their carers – including fitness professionals – to identify people who either have or are at risk of developing an eating disorder, and to prevent or correct the occurrence of the variables that are normally associated with eating disorders.

Key points: Epidemiology and prevalence

- 1 percent of young women suffer anorexia.
- 17 percent of college-aged women suffer bulimia.
- 20 percent of these might die.
- Eating disorders appear mainly in urbanized areas.
- It is mainly women who are affected, but an increasing proportion of male sufferers is reported.

The family of the anorexic[51]

In a landmark study of anorexia nervosa, Mara Selvini Palazzoli, while discussing the family of the anorexic, wrote:

> To the superficial observer, this may look like quite an ideal family. Generally, parents are completely dedicated to their work or to the house, they have a high sense of duty and of social and conventional norms . . . [T]here was, in all cases, a permanent state of underlying tension . . . a marked inclination to endless and unnerving arguments about the most futile issues, which is symptomatic of a hidden aggressiveness which needs an outburst . . . [T]he dominant figure, in the family of the anorexic, is the mother: the father is often emotionally absent . . . secretly or openly underestimated by his wife. Even in cases in which the father, thanks to his intolerant and dictatorial behavior, seems to be the dominant figure, the mother wins . . . stubbornly playing the part of the victim . . . . The daughter easily becomes the victim of the mother . . . the daughter is the ideal baby of an invasive, intolerant and hypercritical mother . . . .[52]

Since Palazzoli's initial observations on the family and its issues, there has been extensive research on eating disorders and family dynamics.[53] Hilde Bruch in her major work, *Eating Disorders: obesity, anorexia nervosa, and the person within*, reports a number of cases that, in her view, shared similar characteristics.[54]

Case history 1

She [the daughter] needed her high class standing, not only for her own peace of mind, but also as an obligation she owed her parents, who, she feared, *would be disappointed if she were not quite so popular and superior*.[55]

Case history 2

Her parents still did not believe that there could possibly be any psychological problems because Christine had been normal and happy to an unusual degree. She was the oldest of four children, had been very helpful with the younger ones, and had been the object of much praise and admiration. She had been a straight-A student, had participated in sports and social activities, and had been popular. It was a shock to see that she hadn't done so well in the college entrance examinations as everybody had expected.[56]

Family dynamics have always been considered as one of the principal factors in the development of eating disorders.[57] Since the earliest studies of eating disorders, psychodynamic descriptions stressed the importance of the family in the triggering and development of the disorder.[58]

Typically, eating disorders appear in middle-class and upper-class families,[59] although some studies have shown that body dissatisfaction, the desire for thinness, and eating disorders are present in other classes as well.[60] These families have internalized the values of the Protestant work ethic (such as success at school and intellectual achievement).[61] They are also characterized by rigidity, enmeshment, intergenerational conflict, over-involvement and avoidance of open confrontation/conflict. They tend to consider obesity as a sign of indolence and lack of willpower, and it has even been suggested that they are often 'sexophobic',[62] i.e. they have a phobia of sex. Families with an eating-disordered member are normally described as highly problematic, manipulative[63] and incapable of deep and stable affective bonds.[64] This brief account of research findings is of course not meant to suggest that families intentionally harm their children, or that children with eating disorders are victims of their families.[65] Research on the family is not meant to stigmatize the family, nor to minimize the drama that many parents and siblings experience. It is meant to detect features that appear common to these families, and over which the sufferer and the family, if willing, could reflect with the assistance of a therapist. According to some experts, given that there are identifiable traits that can understandably generate psychological and emotional distress in all family members, working with the family has a much greater chance of success.[66] A multi-family approach, where a number of families attend psychotherapy together, has also been applied with some success to eating disorders.[67] Fitzgerald and Lane have also stressed the important role of the fathers in the onset of eating disorders. According to these authors, dependency, perfectionism and achievement, which are premorbid traits, are profoundly influenced by fathers.[68]

Another typical trait of these families is the presence of contradictory expectations of their children and, in particular, the daughter. On the one hand, these families value competitive success and achievement (in school and professional life); on the other hand, they also encourage submission instead of the autonomy and independence necessary to obtain the valued success.[69]

This seems to be the background of anorexia and bulimia nervosa sufferers.[70] Although eating disorders are now regarded as multifactorial conditions, that is, as being produced by the interaction of many different causes, research consistently reports that the eating-disordered person has typically suffered abnormal attachment patterns[71] and that family dynamics are intrinsic to both the etiology and treatment of the disorder.[72]

The society

Eating disorders appear nearly exclusively in Western countries[73] or Westernized societies such as South Africa and Santiago (Chile)[74] or countries that are becoming economically emancipated, such as China.[75] For this reason, eating disorders are regarded as 'culturally bound syndromes' – that is, a cluster of signs and symptoms only found in a particular culture or group of cultures.[76] Some experts, for example Mervat Nasser, go as far as to argue that "cultural forces are responsible for this modern morbid phenomenon".[77]

Among the social factors that are thought to be relevant to the spread of eating disorders, we should include those listed in Box 1.2.

Box 1.2 Social factors that are thought to contribute to the spread of eating disorders[78]

1 Centrality of the children in the family: whereas in the past children had a peripheral place within the family, during the twentieth century children have become central to the family. The duties towards children are currently considered very important and being a good/bad parent is a socially discriminatory factor;

2 Longer dependency: the period of dependency of children upon their parent is extended;

3   Greater responsibility of the parents. Points 1 and 2 imply that parents have more responsibilities and for a longer period of time;
4   Change of the social/familial role of the woman;
5   Abundance of food;
6   The social imperative of thinness;[79]
7   The modification of eating habits (culinary multiculturalism; presence of fast-foods; missing lunch; eating alone);
8   Sedentary life and increasing obesity rates.[80]

Many agree that the changes in the role of the woman has had a determinant impact over the way women have perceived their femininity,[81] and consequently on the genesis and spread of eating disorders.

MacSween,[82] for example, argues that eating disorders are somehow due to the modification of the roles of women in society. The confusion and crisis of the anorexic are not due to her own disturbances and defects; it is rather the social world that is lacerated by conflicting expectations about the behavior of adult women. They are expected to be independent achievers, but, at the same time, dependent and willing to embrace their traditional nurturing role.[83] According to Crisp, in order not to face the contradictions of full womanhood, the anorexic refuses food, unconsciously refusing, in this way, to become a woman. She opts for a small body, where smallness is symbolic of a rejection of adulthood, with the conflicting demands it carries with itself.[84] Susie Orbach also stressed how, in modern society, the woman has had to face contradictory demands: on the one hand, autonomy and independence and, on the other, femininity, nurturance, deference and dependency.[85] These roles are incompatible and, according to a societal analysis, the crisis generated by this conflict is expressed by the anorexic through her body which remains trapped in a limbo where the person refuses to feed herself and thus attempts to remain immobile and small when faced with demands that are felt to be impossible to satisfy. Mara Selvini Palazzoli offered a punchy analysis of the demands made of women who are constantly exposed in modern societies to opposing forces:

> Basically, nowadays the woman is asked to be beautiful, elegant and well-kept, and to spend time on her looks; this however, should not prevent her from competing intellectually with men and other women, having a career, and also romantically falling in love with a man, being tender and sweet to him, marrying him and representing the ideal type of lover-wife and oblational mother, ready to give up her degrees . . . to deal with nappies and domestic stuff. It seems evident that the conflict among the many demands . . . represents a difficult challenge for adolescents, especially the most sensitive . . . .[86]

The modification in the social expectations of women is also reflected, according to some studies, in the way women are expected to look. "Looking attractive has always been an important criterion for a woman to feel good about herself and also feel accepted."[87] But it needs to be understood why beauty has been equated with thinness.

Why thinness is beautiful

Studies on the significance of body shape suggest that thinness has historically been valued at times in which women were required to prove their intellectual capacities and that a woman's rounded shape is associated with scarce intelligence.[88]

Vogue, Ladies Home Journal and other magazines have privileged thinness in eras of increased education for women, such as in the 1920s, at the end of the 1960s and during the 1970s.[89] Other studies indicate that winners of Miss America and Miss Sweden pageants, movie actresses, models and, in fact, even Barbie, have dropped several sizes since the 1950s.[90] These changes towards a tubular size seem to reflect the modification of the role of women in society: at the times in which women had to function as men did it was also appropriate for them to look androgynous. Thus their beauty became 'cylindrical' (no curves).[91]

Of course, it is impossible for the vast majority of women to achieve 'manlike' body measures,[92] as the total percentage of essential fat in women, including sex-specific fat, is four times higher than in men.[93] Thus the aesthetic expectations become unrealistic, and this is how, according to this analysis, many women become dissatisfied with themselves: they are expected to look a way they cannot.

Many studies have analyzed the impact of the fashion industry on eating disorders.[94] I will discuss this in greater detail in Chapter 5. Other phenomena, such as 'lookism' and 'healthism', which appear typically in Western and Westernized societies, have also been connected to eating disorders. 'Lookism' is that phenomenon when appearance becomes so important it directly affects people's self-esteem. Sociological studies suggest that those who develop anorexia are fundamentally people who are particularly vulnerable to a culture in which the body is the most direct expression of one's value.[95]

Healthism

Another phenomenon related to eating disorders is that of 'healthism'[96] and other forms of rigid and ritualized control over every aspect of one's life. According to some researchers, eating disorders appear on a continuum with other forms of self-control and discipline. Halse *et al.* have interviewed a number of schoolgirls with eating disorders. The case histories below report two interviews, extracted from their work, which highlight how typically the person with eating disorders has absorbed the belief that self-discipline is a virtue, and should be applied to both intellectual and physical activities.

Case history 3: Elise

At school, [I'll] have a sandwich and maybe a salad and then come home. Have two fruits with – um – and start doing homework. Come five [o'clock], just go for a walk . . . and then I have to – um – have a shower, relax, read and watch TV for half an hour. Then have dinner, start my homework again and come nine o'clock, I'll go skipping for 15 minutes.[97]

Case history 4: Laura

As Laura explained, her daughter would:

. . . go and get a packet of frozen vegetables or a frozen dinner or whatever out. She would read the nutritional value on the back. She would write it down. She would pour it out of the bag. She would still put it on the scales and weigh it, then heat it, and take it out, and reheat it, and weigh it and put it down. And she would eat it in the same order. And it was every single meal, every single day.[98]

One objection to the results of these sociological studies is that anorexia is not a phenomenon of modern times. Sociological studies argue that eating disorders are a distinctly modern condition caused by the social pressures discussed above. However, as we saw at the beginning of this chapter, anorexia was first conceptualized as a disorder in the nineteenth century. The history of self-starvation is even older than that. Fasting has always been used as a form of purification in religious rituals, and as a form of political protest. In modern times it has been thought of as a mental illness. Vandereycken and Van Deth[99] have asked whether those forms of self-starvation could be comparable to anorexia and eating disorders. They have offered a very interesting and thorough study of the similarities and differences between different forms of self-starvation. They point out that the types of self-starvation that we find in human history before the nineteenth century present important differences to the starvation that characterizes eating disorders. Eating disorders cannot be fully equated to the more ancient self-inflicted harm that characterizes ascetic practices. I refer the reader to Vandereycken and Van Deth's study for more. But I accept their conclusion that anorexia is not a type of self-inflicted starvation that has *just been conceptualized* as a mental illness in our era. The forms of fasting that characterize asceticism, or political protests, have notable differences. Anorexia and eating disorders generally are not properly comparable to these other phenomena of self-inflicted harm.

In other works I have discussed the results of sociological analyses of eating disorders, and have highlighted their strengths and weaknesses.[100] This brief outline offered here should be enough to equip the reader with a sufficiently broad understanding of what eating disorders are, and who is most at risk. Results of studies on the family and society of the eating-disorder sufferer are, in fact, important not only because they enhance our understanding of the condition but also because they give us some ways to identify a person at risk of developing an eating disorder. This person is most likely to be a woman, and most likely to be of Western origin. She is most likely part of an upper-middle class family, and she will normally be intelligent and well educated. Moreover, she will not be 'just thin'; her exercise regime, like her whole life, will be rigid, tightly regulated and fully or highly ritualized.

Conclusions

This chapter has offered a description of eating disorders, based on case histories and on sociological literature. Incidence, prevalence and mortality rates have been reported, as well as the features that, in clinical psychology, are found to be common to those who are afflicted by eating disorders. Of course, there is more to eating disorders than the characteristics highlighted in this chapter: there is a real person, who fitness professionals and leaders are called to assist in the sports and fitness environment. The language of clinical psychology might at times describe people's experiences as 'symptoms' and 'signs' of an illness. I have deliberately talked about eating disorders as a 'condition' rather than an 'illness'. As will become clearer in Chapter 9, there is no reason to consider eating disorders as a mental illness, and whether or not people prefer conceptualizing eating disorders as mental illnesses is irrelevant, from an ethical point of view. How we describe what happens to the sufferer says nothing about how we should deal with them. Eating disorders in this book are regarded as an existential condition, a series of experiences that happen to the person. I will discuss whether there are physiological or genetic factors that might explain why these experiences occur. The focus on clinical studies and research findings is not meant to dehumanize the experience of sufferers: clinical research is relevant to our purposes because it helps us to understand better what the sufferers' experiences are; we can appreciate the psychological, familial, emotional and physical sequels associated with their condition; and we can also discuss whether it can be ethical to intervene against the sufferers' wishes, and how, for their own good. This, in the context of exercise and fitness, should not detract from, but should eventually add sensitivity and consideration to, the relationship with the eating-disordered exerciser.

The next chapter summarizes the threats that eating disorders pose to our health.

Key points

- A person with an eating disorder is at serious risk, even if at normal weight.
- Anorexia involves continuous fights against hunger; bulimia involves overeating followed by purging (especially self-induced vomiting).
- Binge-eating disorder and obesity may also be included among eating disorders.
- Anorexia athletica is the condition of athletes who adopt disordered eating patterns.
- Exercise is used by about 50 percent of women with eating disorders to control body weight.
- Bulimia can be more difficult to detect than anorexia.
- Bulimia can be as life-threatening as anorexia.
- Exercise is a threat for the eating disorders sufferer, even at normal weight.
- Absence of menstruation can cause bone thinning and increases the risk of fractures during exercise.
- Eating disorders appear nearly exclusively in Western or Westernized societies.

Although identifying a person at risk of having, or developing, eating disorders is a great challenge for fitness and sports professionals, this chapter has provided data that can help them to detect people at risk. The following chapters will provide more of these summary boxes.

Box 1.3 Identification kit for fitness professionals

- Irregular or absent periods could be a sign of an eating disorder.
- Recurrent cramps could be a sign of eating disorders.
- Women are at higher risk of developing eating disorders.
- The eating disorders sufferer will be most likely of Western origin (although people of a different ethnic background might also be affected).
- Eating disorders sufferers are more likely to come from middle- and upper-class families (although people belonging to different social classes might also be affected).

2　The effects of abnormal eating

Introduction[1]

Abnormal eating is a serious threat to people's health. Mortality associated with eating disorders is "20 percent at 20 years [. . .] by far the highest of any functional psychiatric illness, and high compared to most chronic medical illnesses".[2] Other studies report 20 percent crude mortality for anorexia and bulimia nervosa after a mean of 12.5 years of follow up.[3] However, the consequences of food restriction and compensatory behavior may be severe and life-threatening, and this is so even if the person is not extremely thin.

This chapter discusses the effects of eating disorders on the endocrine, cardio-vascular and gastrointestinal systems. I will show that the 'secondary symptomato-logy' ('secondary' means that the symptoms are not the primary disorder, but the effect of abnormal eating) is usually reversible. This means that it generally disappears when normal eating is re-established.

Effects of eating disorders on the endocrine system

The endocrine system consists of glands (pituitary, thyroid and adrenal) which secrete growth hormones, prolactine and other hormones that regulate physio-logical functions such as metabolism.

Eating disorders affect the endocrine system in a significant way.[4] Typically women stop menstruating (amenorrhea) or develop irregular menstruation cycles (dysmenorrhea).[5] The eating disorder may be a cause of fertility problems in adult women. Males may suffer from impotence and lack of sexual interest (low libido). Sometimes growth of skin hair (lanugo) also occurs due to hormonal imbalances.

Amenorrhea is one of the signs of a decreased level of estrogens.[6] Estrogens are hormones that are also responsible for the health of our bones. Decreased levels of estrogens cause bone weakening.[7] Also men with eating disorders are at risk of osteoporosis.[8] Loss of bone mineral density obviously increases the risk of fractures during exercise[9] and of long-term osteoporosis.[10] It should be noticed that hormone replacement does not compensate for the natural loss of bone mineral density.[11] Lyn explains:

Estrogen administration alone has not been shown to prevent progressive bone loss in current diagnosed anorexics and increases in weight alone appear to be insufficient in reversing bone mineral density losses. [. . .] Even young women who recover before 15 years of age have been shown to have long-term decreased bone density in the lumbar spine and femoral neck.[12]

The absence of menstruation is important for both fitness instructors and sufferers who exercise because of an associated risk of fracture. An early onset eating disorder such as anorexia nervosa may block normal development in puberty leading to growth retardation.[13] Overexercising or excessive physical training in children may have a similar impact on growth.[14]

Cardiovascular disorders

Poor nutrition also has detrimental effects on the heart. Like any other muscle in the body, the heart can increase or decrease in size, and become stronger or weaker. Typical effects of extreme diet are low heart rate (bradycardia) and low blood pressure (hypotension). The heart becomes smaller and weaker. The heart, by pumping blood, delivers oxygen to the tissues, including the muscles. While exercising, the working muscles require more oxygen and more blood to function effectively, and the heart needs to pump more blood to supply extra oxygen. Since the heart of the person with eating disorders is generally smaller and weaker, its ability to deliver oxygen to the tissues while exercising may also be impaired, with potentially dangerous consequences. According to some researchers, the mitral[15] valve may prolapse potentially causing fatal arrhythmias.[16]

Electrolyte disorders

Anorexia and bulimia nervosa also cause electrolyte disorders. Electrolytes are substances capable of conducting electricity. The most important are sodium, potassium and chloride. Vomiting and (ab)use of laxatives cause dehydration and electrolyte imbalance. The most common result of this is cramps (especially due to lack of ATP – adenosine triphosphate, which is one of the carriers of energy in human cells[17] – and low potassium level).[18] However, electrolyte imbalance might also cause epileptic attacks, severe abnormalities in heart rhythms, respiratory paralysis,[19] cardiac arrest and death.[20]

One major problem which especially affects fitness professionals is that individuals with low potassium levels are often asymptomatic, and it is therefore impossible to predict when a potentially life-threatening cardiac arrhythmia may occur.[21]

Gastrointestinal disorders

Gastrointestinal disorders are another typical consequence of low and abnormal nutrition. These include swollen sub-mandibular glands and parotid glands, esophagitis, esophageal spasm and esophageal tearing and potentially fatal ruptures. This may occur from frequent vomiting. Bulimics may also suffer from undiagnosed gastric and esophageal disorders which may contribute to involuntary reflux and vomiting.[22] Bingeing may cause gastric dilatation.

> Acute pancreatitis can occur as a result of binge eating or in anorexics who are re-fed. Long-term laxative abuse can cause pancreatic damage and inhibit normal insulin release. Atrophy of the pancreas has been observed in anorexics and, although the size of the pancreas appears to revert to normal with recovery and increases in body weight, it is unknown whether pancreatic function returns to normal. Abnormal motility, reflected in delayed gastric emptying, increased transit time, constipation, loss of peristalsis, irritable bowel syndrome, steatorrhea, and melanosis coli (a dark brown discoloration of the colon secondary to laxative abuse) can have a variety of causes including binge eating, purging, food restriction, laxative abuse, electrolyte deficiency (potassium, magnesium) and dehydration.[23]

Repeated vomiting can dissolve tooth enamel and cause recession of gums, and binge eating can cause expansion and even rupture of the stomach.

Conclusions

This chapter has discussed the main effects of abnormal eating on the endocrine, cardiovascular and gastrointestinal systems. The consequences of abnormal nutrition on the body are of clear relevance to anyone treating the eating-disordered person, but in the fitness and sports environment they acquire special importance, as physical activity can become an obvious additional threat for the sufferer. Even when emaciation is not advanced, the person with eating disorders is typically debilitated. The design of physical activity needs to take into consideration the strain that abnormal eating poses on the sufferer's body. Moreover, the fitness professional needs to be aware of not only how thin a person is but also of their body mass index and other measures of body fat or weight. The fitness professional might also need to be alert to other signs of eating disorders because people with eating disorders are not always emaciated and might have normal body weight – yet their physique might be debilitated.

Key points

- Eating disorders affect the endocrine system, causing amenorrhea and other hormonal imbalances.
- Amenorrhea increases the risk of fractures during exercise and long-term danger of osteoporosis.
- Eating disorders affect the heart.

- Eating disorders can cause cramps, epileptic attacks, severe abnormalities in heart rhythms, respiratory paralysis, cardiac arrest and death.
- Eating disorders can cause gastrointestinal disorders, such as esophagitis, constipation and irritable bowel syndrome.

3 Biological and clinical explanations of eating disorders[1]

Introduction

Understanding the causes of eating disorders is important to determine the best course of action with eating-disordered exercisers. If eating disorders were a biological illness, or an addiction, it would be imperative to involve qualified and specialized medical personnel. If they were mainly the result of free choice, this would mean that, at least in principle, negotiation with the sufferer is possible. Little negotiation, if any, instead, is possible with someone affected by a neurological or genetic disorder. If an Alzheimer's patient is carelessly crossing a busy road with incoming traffic while in a confused state, there is limited scope for dialogue with the sufferer. She cannot help it. Where self-harm cannot be explained exclusively in terms of determining biological factors, this indicates that the person retains the power to choose, on some level at least, that this is the behavior s/he will adopt.

Those who have the strength of will to inflict so much pain on themselves should have, and often do have, enough strength to resolve the conflicts that lead them to fight against their own self. For these reasons, understanding the causes of eating disorders is ethically relevant. It puts the condition into context, and helps us to find out whether paternalism is ethically mandatory or whether, instead, there is sufficient scope for meaningful dialogue.

This and the next two chapters will therefore discuss the various possible causes of eating disorders. This chapter assesses whether eating disorders are a biologically determined illness, the next chapter whether they are forms of addiction and Chapter 5 whether they are caused by the pressure, often operated by the media, to be thin.

Box 3.1 Dictionary definitions, based on those in the *Oxford English Dictionary* (available online at www.oed.com)

Biochemistry: The science dealing with the chemical substances present in living organisms and with their relation to each other and to the life of the organism; biological or physiological chemistry.

Biology: The division of physical science that deals with organized living animals and plants, their morphology, physiology, origin and distribution. There are multiple branches of biology: for example, genetics, neurology, psychology, molecular biology, etc.

Chemistry: That branch of physical science which deals with the elementary substances, or forms of matter, of which all bodies are composed, the laws that regulate the combination of these elements in the formation of compound bodies, and the various phenomena that accompany their exposure to diverse physical conditions.

Endocrine system: Denoting a gland having an internal secretion that is poured into blood or lymph; a ductless gland such as the thyroid, pituitary, and adrenal glands.

Physiology: A branch of biology that studies the normal functions and phenomena of living things. It comprises the two divisions of *animal* and *vegetable (plant) physiology*; that part of the former that refers specially to the vital functions in man is called *human physiology*.

Neuroendocrinology: The study of the interactions between the nervous system and the endocrine system.

Neurophysiology: The study of the physiology of the nervous system.

Are eating disorders a genetic disorder?

Eating disorders are thought to result from a complex interplay between environmental and genetic[2] risk factors. When a disorder results from many different causes, in scientific language it is said that *its etiology is heterogeneous*.[3] Diseases with heterogeneous etiology are known as *complex or multifactorial diseases*.[4] It seems that eating disorders must be explained in a multifactorial manner.

There are two types of multifactorial diseases: in one type, only a small number of genes behave differently from normality (gene variants). In these cases the genetic component is said to be *oligogenic*. In the other type, there are many gene variants that act simultaneously and interact: here the genetic component is said to be *polygenic*.[5]

Box 3.2 Types of genetic diseases

Multifactorial or complex diseases: Disorders resulting from many different causes; or, disorders whose etiology is heterogeneous.

Gene variants or susceptibility alleles: Genes that behave differently from the norm.

> **Oligogenic diseases**: Diseases where only a small number of gene variants is present.
>
> **Polygenic diseases**: Diseases where many gene variants act simultaneously and interact.

It is unclear whether eating disorders might have an oligogenic or polygenic genetic component, because, as we are going to see, studies on the genetics of eating disorders have not yet provided conclusive answers.

Many healthy people in the general population may present the genetic variants that contribute to complex diseases (both oligogenic and polygenic), but do not develop the diseases for which they carry the genes. This means that these gene variants are not necessarily deleterious and it is not certain that they *will cause* the disease.[6]

So, what causes *some* individuals to develop certain multifactorial diseases, if the gene variants alone are not sufficient to cause them?

Genes and environment: A mysterious interplay

Despite the fact that genetic abnormalities are not solely responsible for eating disorders, many scientists seem to agree that genetic predisposition plays an important role in the development of eating disorders.[7] Yet the modality of interaction between genetic and environmental factors, and the extent to which genetic factors are involved in susceptibility to eating disorders, are unresolved issues.[8] It is likely that the interplay between genetics, environmental influences, and the individual's interpretation of these factors in one person's articulation of behavior is very difficult, if not impossible, to capture.[9]

Anorexia and bulimia nervosa are found to be "statistically more common among family members"[10] than in the general population. Studies on twins have shown a concomitance of 50 percent between monozygotic (or identical) twins, as compared with 10 percent of dizygotic twins.[11] This concordance is called the 'concordance rate'. Dizygotic twins are those twins who have different DNA; they are not identical, but happen to share the womb at the same time. The greater concordance rates among monozygotic twins as compared to the concordance rates among dizygotic twins is generally taken as evidence of a "strong etiological role for genetic factors".[12]

In other words, among identical twins, it is quite likely that if one has an eating disorder, the other has it too. Among non-identical twins, even if they are also raised in the same environment and born at the same time, it is five times less likely that both of them will develop an eating disorder. This seems to indicate that, since identical twins have the same DNA and non-identical twins do not, genetics is somehow responsible for eating disorders.

However, the greater concordance among identical twins does not necessarily mean that genetics is the cause of eating disorders. To render these results more reliable, it would be necessary to study identical twins brought up separately, and to assess whether there is also a concordance rate among identical twins brought up separately, and what this is. Studying separated twins would give us more precise information on the effective influence of genetics on individuals' behavior and predisposition to diseases. However, to date, there have been no adoption studies of eating disorders among separated monozygotic twins.[13]

There are also methodological difficulties with case ascertainment, and scientists recognize the low statistical power of available studies.[14]

The interpretation of familiality (incidence within a family) is thus not straightforward.[15] Some studies "strongly suggest that the familiality observed in family studies is primarily due to genetic causes";[16] other studies suggest that this familiality is likely to result from both environmental and genetic influences.[17] Some researchers argue that genetics is the most determining factor – or that there is a genetic predisposition to anorexia, which becomes manifest due to environmental stressors, such as inappropriate diet or emotional distress;[18] other researchers stress the importance of environmental stressors – and argue that environmental influences play a major role in determining the onset of eating disorders.[19]

Some researchers have noticed that genetic explanations cannot capture a number of aspects of eating disorders. They argue that *non-shared environmental experiences* are also significant in the genesis of the disorder. Non-shared environmental experiences are those unique to each individual, despite the fact that they live in the same family. For example, two siblings may share the same genetic make-up, the same family and the same environment, but while some experiences will be similar, other experiences and influences will be *non-shared*, that is, unique to each individual. Monozygotic and dizygotic twin studies can thus also be used to understand the importance of non-shared environmental experiences in the development of eating disorders.[20]

In short, genes might be involved in the onset of eating disorders. However, genes do not seem to be solely responsible for these conditions. In other conditions, genes are to a very large extent, or completely, responsible for the symptoms. Tay-Sachs disease, for example, is a neurodegenerative disorder entirely caused by mutations in the HEXA gene. Muscular dystrophy is another example of a genetic disorder where the environmental stressors might play a role in retarding or containing the degeneration, but which is largely determined by genetic mutations. The case of eating disorders is different from primarily genetic disorders. In eating disorders, genetics, in interplay with environmental factors, individual sensitivity and social pressure, contributes, in a somehow mysterious way, to generate these still scarcely understood conditions.

Other researchers, rather than focusing on the genes,[21] have suggested that eating disorders are caused by dysfunctions in the central nervous system. It is to this suggestion that I now move.

The brain: The central nervous system

There seems to be evidence that people with eating disorders have altered brain structures. In some cases, the alteration is still present after the disorder in eating is corrected. In view of these alterations and of the symptoms of eating disorders, some experts have suggested that anorexia and bulimia nervosa may be caused by a disorder in the neurotransmitter system.[22] Neurotransmitters are elements present in our central nervous system, which are responsible for the transmission of signals within the brain. Some of these neurotransmitters seem altered in eating-disordered patients.[23]

In particular, dopamine is responsible for giving us sense of hunger and satiety, and therefore some scientists think that it may be involved in eating disorders.[24] Endorphins and serotonin are generally associated with states of well-being and excitement. In eating-disordered people, the levels of these neurotransmitters are unbalanced.

It has also been suggested that opioids may play a role in generating eating disorders. Opioids influence eating: they increase the desire to eat, whereas opioid antagonists decrease it.[25] Several researchers have noticed that the level of opioids in both anorexics and bulimics is higher than normal.[26]

The relationship between the change in the rate of opioids, dopamine and eating disorders is, however, unclear.[27] Although there is an association between eating disorders and imbalances in the neurotransmitter system, many of these imbalances seem to be the consequence of eating disorders, and tend to return to normal levels as weight is gained.[28] Abnormalities in the level of serotonin may in some cases persist after weight gain and it has therefore been hypothesized that a disorder in serotonin may create a vulnerability to anorexia and bulimia.[29] The causality, however, has not been established.

The central nervous system thus remains a fertile field of investigation for research into the causes of eating disorders, but research is still ongoing.

The hypothalamus

The hypothalamus is the portion of the brain that secretes substances that control metabolism, nutrition,[30] reproductive activity, temperature, homeostasis, sleeping and wakefulness.

The hypothalamus has three parts, or axes: the *hypothalamo-pituitary-thyroid axis*,[31] the *hypothalamo-pituitary-gonadal axis* and the *hypothalamo-pituitary-adrenal axis*[32] (Box 3.3). Eating-disordered people show variations in all areas regulated by the hypothalamus: they are normally intolerant to cold, suffer constipation, dry skin and hair, low heart rates, slowly relaxing reflexes and low metabolism rate. Also, anorexic women typically suffer amenorrhea, whereas males suffer hypogonadism, low libido or difficulty with erection. Studies[33] on eating-disordered people have also shown that people with anorexia and bulimia present thyroidal dysfunctions.[34] For more information on the hypothalamus and its relation to eating disorders, see Box 3.3.

Importantly, similar imbalances are not found in people who have lost weight because of physical illnesses, and therefore these imbalances do not depend only on weight loss.[35] However, these abnormalities are corrected when normal weight is restored[36] or when the person interrupts compensatory practices such as self-induced vomiting and use of purgatives and diuretics.[37]

It is important to note that there is one partial exception to this:[38] in anorexic females, weight restoration does not automatically lead to the onset or resumption of menstruation. This could be because the female reproductive system is highly sensitive to psychological or physical stress.[39] This, as I shall argue in Chapter 6, is relevant to fitness practice.

In any case, the neuroendocrinological system and the biological processes regulating nutrition in humans are not fully understood,[40] and therefore it cannot be claimed that disorders in the hypothalamus are a cause of eating disorders.[41] As the imbalances relating to the hypothalamus are reversed as normal nutrition is restored, it is likely that eating disorders are the cause of dysfunctions in this area of the brain.

Box 3.3 The hypothalamus and eating disorders

Hypothalamus	*Eating disorders symptoms*
Hypothalamo-pituitary-thyroid axis Regulates tolerance to cold, intestinal functioning, skin hydration, heart rate, reflexes, metabolism.	Eating-disordered people are often intolerant to cold, suffer constipation, dry skin and hair, low heart rates, slow-relaxing reflexes and low metabolism rate.
Hypothalamo-pituitary-gonadal axis Regulates reproductive functions and libido.	Amenorrhea in females with eating disorders; hypogonadism often found in males, with low libido or difficulty with erection.
Hypothalamo-pituitary-adrenal axis Regulates reactions to stress, digestion, mood and the immune system.	Eating-disorder sufferers have imbalances in these areas. Similar imbalances are not found in people who have lost weight because of physical illnesses.

Conclusions

A scientific understanding of eating disorders provides important data which are relevant to fitness professionals. Eating disorders are a serious health problem. They affect the brain and whole organism. Their gravity should never be underestimated. Insofar as serious medical problems result from abnormal eating, fitness professionals should always consult more qualified colleagues or medics, in order to decide whether exercise can be safe for the individual, and how to design a proper training program for the eating-disordered exerciser.

There is no reason to consider the symptoms of eating disorders solely as the result of genetic or biological disorders. Therefore, in principle, sufferers should be able to control what happens to them, and negotiate a healthy exercise program with their exercise instructors. However, establishing a relationship of trust with an eating disorders sufferer is notoriously difficult. Sufferers are often very shy and embarrassed about what is happening, and are reluctant to talk openly about how they feel, or about their eating habits. They protect strenuously and at all costs those rituals which, bit by bit and with meticulous observance, are eroding their own body and alienating them from the rest of the world. Eating disorders sufferers industriously excavate a void inside themselves and, as they do so, they create a gulf of misunderstanding between themselves and other people. The person with anorexia empties herself and so becomes distant, unreachable, unable to listen and even perceive the pain, despair and fear of others. It is therefore typical for a sufferer to first seek assistance or help, and then retreat and refuse advice.

Sufferers and carers have a sense of dealing with an illness that 'possesses' the sufferer. Sometimes both carers and sufferers describe eating disorders as an addiction, as an illness that takes over and leaves the sufferer with no willful control over her/his behavior. Carers often have the sense that the disorder, like heroin or other psychoactive substances taken habitually, has changed the personality of the sufferer. In the next chapter, I will discuss the issue of whether eating disorders can be regarded as a form of addiction.

If the addiction hypothesis turns out to be valid, this would have implications for fitness practice. First, it would mean that establishing a relationship of trust and negotiation with the participant would be not only difficult but nigh on impossible, as the client is not free to choose but, instead, is driven by a physical impulse that has a firm neurophysiological basis beyond the sufferer's control. Second, it would mean that the fitness professional is facing a task that goes probably beyond the competencies that s/he has as a fitness professional. Finally if eating disorders and over-exercise are addictions, this would have important implications regarding the ethics of coercive intervention toward sufferers. In principle, paternalistic interference would be justified. This would apply to the fitness arena. The issue of respecting the autonomous wishes of the client would be less persuasive, insofar as these wishes were proven to lack autonomy in some relevant way (see Chapter 9).

Key points

- Eating disorders are complex or multifactorial diseases.
- Genetics seem to be involved in the development of eating disorders.
- Genes and environment both seem to contribute to the onset of eating disorders.
- Eating disorders affect the brain.
- Most alterations of the central nervous system are normalized when eating-disordered patterns are abandoned, and normal weight is restored.
- Some physical alterations persist, even after recovery.

4 Eating disorders, exercise and addiction

The Gus Goose syndrome

Some of us will remember Gus Goose,[1] one of Walt Disney's characters, cousin of Donald Duck and grandnephew of Grandma Duck. He lives with Grandma Duck, and is supposed to help her out with the farm. But it is extremely hard to distract Gus Goose from his own favorite activities: eating and sleeping. Gus Goose is the fictional representation of gluttony and laziness. He is probably one of the most famously addicted characters of the history of cartoons: addicted to food and sleep. Gus Goose can be regarded the worst nightmare of eating disorders sufferers – he is what they do not want to become. Those with bulimia, in particular, might recognize themselves in Gus Goose, incapable of resisting food and able to ingest enormous amounts . . . but while Gus Goose is perfectly happy with food, and with his rounded bottom half, bulimics feel miserable if they overeat, and are very unhappy with the way they look, even if they are not fat.

Is there a 'Gus Goose Syndrome'? Can food (or food restriction) generate addiction? Understanding whether eating disorders and over-exercise are forms of addiction is central to our understanding of how we should approach the eating-disordered exerciser. If these abuses are forms of addiction, this would give specificity to the diagnosis as well as to treatment. It would mean that eating disorders should be managed by qualified health care professionals specialized in disorders of dependence, in the same way as other potentially lethal addictions, such as heroin addiction or alcoholism. Moreover, from an ethical point of view, as addictions jeopardize decision-making capacity, restrictions of freedom would be, in principle, justified (see Chapter 9, 'Weak paternalism'). If eating disorders and exercise are not addictions it would mean that the person plays an important role in shaping her/his behavior. Paternalism, in this case, would be much harder to defend, and could not be justified on the grounds that the person is lacking capacity 'because s/he has an addiction'. It would probably also mean that fitness professionals

Figure 4.1 Gus Goose.

are called to make a much more complex judgment as to how to direct the relationship with these vulnerable clients, because it might not be true that eating-disordered exercisers do not have control over their behavior, but instead there might be a profound ambivalence in the sufferer, a strong need to feel well with oneself, and a strong need to harm oneself. It is probably this ambivalence that makes it so hard to establish a meaningful management-plan with eating disorders sufferers. In this chapter I argue that eating disorders and over-exercise cannot be considered as addictions.

The Addiction Model

As we are now going to see, many have tried to demonstrate that eating disorders and over-exercise are forms of addiction. The Addiction Model is a general term that refers to various theories, which have one point in common: they all consider an addiction, or something similar to a substance use dependence, as essential to the etiology of eating disorders and over-exercise.[2] Some authors have linked eating disorders to obsessive-compulsive disorders,[3] claiming that the fear of fat (obsession) leads the sufferer to adopt repetitive and potentially self-harming behaviors (compulsion).[4] Addictions, or substance-use dependences, are variants of impulse control disorders. In disorders of dependence, the person is motivated by an impulse, which is a drive towards a pleasurable, or desirable, outcome (for example, the desirable feeling from a psychoactive drug or even of a cigarette). In *compulsions*, instead, the repetitive action is intended to avoid a negative, or undesirable, outcome (for example, the urge to wash hands as a response to the terror of feeling dirty). The demarcation line between these two compelling states, between impulsive and compulsive behaviors, is not clear. Even ordinary smokers, for example, know that they will resort to a cigarette to cope with anxiety or other negative emotions. People, in other words, also use the substance of addiction to release negative emotional states. The bingeing behavior lies between impulsion and compulsion. In some sense, bingeing is impulsive, as it is experienced as uncontrollable and providing pleasure, but it is also typically an anxiety-reducing compulsive behavior. Many sufferers binge when they are tense or anxious. In the Addiction Model a general core idea is that people do not really want to adopt abnormal eating and exercise patterns, but are *compelled* to do so. Their behavior is *out of their own willful control*.[5] The following case history is an illustration of this core idea.

Case History 1[6]

Melissa has been an athletics track fanatic since she was 12 years old. She has run the mile in meets in junior high and high school, constantly improving her times and winning several medals. Best of all, Melissa truly loves her sport.

Recently, however, Melissa's parents have noticed a change in their daughter. She used to return tired but happy from practice and would relax with her family, but now she's hardly home for 15 minutes before she heads out for another run on her own. On many days, she gets up to run before school as well. When she's unable to squeeze in her extra runs, she becomes irritable and anxious. And she no longer talks about how much fun track is, just how many miles she has to run today and how many more she should run tomorrow.

Melissa is living proof that even though exercise has many positive benefits, too much can be harmful. Teens, like Melissa, who exercise compulsively are at risk from both physical and psychological problems.

This story illustrates a typical case of over-exercise. Moreover Melissa's behavioral changes are attributed to exercise: 'too much can be harmful'. In the Introduction to this book, we have seen that Christy Henrich's mother said that "99 percent of what has happened to Christy is because of the sport". Christy, a gymnast with anorexia nervosa, died at the age of 22.

The message in these claims is that exercise is potentially dangerous: if you exercise too much, exercise can 'take over', can become addictive, and you might become unable to decide if, when and for how long you will exercise. A similar explanation is sometimes given for eating disorders. The idea is that diet can 'get out of control', and that the dieter, like the exerciser, can develop an addiction to unhealthy eating behaviors. Let us now try to understand whether sport, fitness or physical activities, and diet can be addictive.

The idea of addiction surely captures the experience of the sufferer who often feels unable to modify her or his behavior in spite of self-harm, and the experience of carers who struggle against a wall of irrationality and incommunicability. They witness their loved ones slowly destroying themselves, and lack the power to help them. First, we should ask: what is 'an addiction'?

Addiction: Two meanings

Literally, addiction has several meanings. The two most pertinent meanings to our discourse, as given by the Oxford English Dictionary, are the following:

1   The state of being (self-)addicted or given to a habit or pursuit; devotion.
2   The, or a, state of being addicted to a drug; a compulsion and need to continue taking a drug as a result of taking it in the past.

Both these definitions are circular and beg the question: addiction is defined in terms as a state of addiction. This lack of precision in the definitions is not important to our discourse. What we need to establish is whether eating disorders can be a form of devotion to a pursuit, or, instead, a stronger compulsion, possibly explained by a physiological process.

I will call the first meaning a 'weak meaning' of addiction, and the second a 'strong meaning'.

In the weak sense, 'addiction' refers to *a strong desire or inclination to do something*. In the strong sense 'addiction' refers to *a dependence*, generally on a psychoactive substance. Addiction in this sense is an *impulse towards* something. In this second sense addiction is rooted, at least to an important degree, in neuro-physiological processes. The repeated assimilation of psychoactive substances alters the release, storage and natural transit of neurotransmitters, and reduces the capacity of the brain receptors to produce those substances endogenously. Once the substance is withdrawn, the brain is unable to restore the normal supply, and this causes the withdrawal syndrome.[7] The signs of this withdrawal syndrome (e.g. sweating and trembling) will disappear after the intake of the psychoactive substance (e.g. alcohol). This is seen as a proof of real dependence or addiction in the neurophysiological meaning of the term.[8]

Thus, in everyday language, it is equally appropriate to say that I am 'addicted to shopping'; or 'addicted to my son' because 'I can't stop cuddling him', and that 'Jeff is addicted to heroin', although these sentences all have different meanings. Heroin is, in fact, a psycho-addictive substance – my son is not. My son stimulates the endogenous production of chemicals that give me pleasure.[9] Heroin, on the contrary, jeopardizes and damages the endogenous production of these chemicals. In ordinary discourse, these two meanings are often confused, but the literature on eating disorders, exercise and addiction also often reflects the flexible meaning of addiction in everyday language.

Part of the ambiguity relates to the fact that everything pleasurable (or unpleasant) produces 'physical' variations, and no emotion is skimmed from chemical, biological and neurological components. This means that the demarcation line between addiction in the weak and in the strong sense is naturally blurred.

In spite of this, there is a difference between addictions in the strong and in the weak sense, and this difference should be retained as clearly as possible, in order to understand whether eating disorders and exercise are forms of addiction, and, eventually, in what sense.

The weak meaning of addiction

Eating disorders and exercise cannot and should not be considered addictions in the weak sense of the term. The weak definition of addiction is all-inclusive and vague, and risks presenting all behavior as pathological.[10] If addiction is used in a 'weak' sense, in fact, it follows that we are all addicted to many things that we like, or love, and without which we cannot imagine our life. A large part of our behavior is dictated by pleasure or positive rewards. Some behaviors may be potentially very harmful (some jobs, or some sports, for example, are highly risky), but some might not, and yet we cannot separate ourselves from them. The paradox of this model is thus threefold:

First, under this model, normal experiences of desire, pleasure, devotion, pursuit, passion, are pathologized regardless of whether we value those experiences and consider them as contributing to the flourishing of our life.

Second, those who we normally admire because they take risky jobs or activities for devotion to ideals or values would be regarded as addicted, whereas those who undertake risky jobs out of financial need or scarcity of alternatives would not be considered as addicted – a counter-intuitive incongruence. The 'highly committed' might be regarded as 'addicted' under this definition.[11] For example, a devoted journalist who offers to report from dangerous areas where few others are willing to travel because he believes that the world should know what's going on in these areas will be regarded as 'pathological'. His choice, under this model, is not free in an important sense because it is motivated by devotion and strong desires, and is performed at high personal cost, whereas a man who is forced to work in an unhealthy environment because of extreme need (a need that leaves him with no choice) would be instead regarded as 'non-addicted', because he is moved by neither love nor passion for his work, nor choice. The paradox here is evident. It would be reasonable to regard the worker as not being as free as the journalist.[12]

If we use 'addiction' in this weak sense, either all behaviors driven by devotion or associated with strong pleasure or desire are pathological, or it is unclear why eating disorders should be considered a problem at all. Abnormal eating and exercise would be just one of the many 'addictions' that we all have. Thus, the third paradox is that unless one pathologizes every behavior (or every risky behavior) driven by devotion, desire or love, eating disorders and other abuses of the body are normalized under this model.

It is obvious that even behaviors that are freely chosen are also to some extent determined (by genetic constitution and by other given circumstances beyond our willful control).[13] For example, whether I excel in tennis or figure skating will depend, to an important extent, on my genetic constitution. Therefore my choice to participate in figure skating is partly determined by my genetic make-up. However, much more than biological or genetic predisposition is involved in what we do – and how, when, in what circumstances and with whom we do it. A constellation of reasons and motives leads us to behave in a certain way, or to devote energy to the pursuit of some goals rather than others. The fact that, to an important extent, our biological or genetic constitution equips us better for some tasks rather than others, and the fact that sometimes we truly love what we do, does not justify regarding our activities as pathological. Having desires that persist over time, preferences that induce us to shape our lives in certain ways, values or hobbies to which we are devoted, is all part of life and, probably, of good life. Instead, it is reasonable to consider eating disorders and over-exercise as behaviors of peculiar significance, and of particular concern; behaviors that need to be understood and hopefully prevented or remedied.

One could say: an addiction is a form of devotion *to something that is harmful*. The weak meaning is correct, but needs to be modified. Thus, for example, being devoted to a good partner is not a form of addiction. Having a good habit, for example a good diet, is not an addiction. Instead, being devoted to a destructive relationship, or starving yourself, might be a form of addiction. What makes devotion an addiction is the consequences of that devotion. The harm could be emotional, social or financial. Enjoying gambling might not be an addiction: it becomes an addiction when the social and financial consequences are bad enough. Yet, it cannot be the harm that turns devotion into addiction. The journalist, in the example above, might be harmed, and might even risk his life. Yet, it would be unreasonable to regard the pursuit of his goals, which are worthwhile for him, as addictive. If this were the case, again, many types of pursuit and devotion would be regarded as forms of addiction. The weak version of addiction, even if modified on the basis of harm, is still too vague and all-encompassing. Moreover, harm is incidental and circumstantial. How much harm I suffer from, for example, gambling, or smoking, is also a function of my financial circumstances or genetic make-up. Harm, thus, cannot be what makes devotion and habits addictive. There must be something else to addiction.

One could then say: the core meaning of addiction is being devoted to something harmful, and continuing to pursue it, not only in spite of harm but also in spite of a strong wish *not to be devoted* to that thing. The core feature of addiction, in this sense, would be the inability to modify one's devotions and one's behaviors. Thus, what makes eating disorders (and over-exercise) addictions is the fact that they are habits, that they are *harmful*, and that the person *wants to quit them, but cannot.*

Bringing these further specifications into the weak meaning of addiction still does not seem to prove that eating disorders and over-exercise are forms of addiction. First of all, it is not clear that people with eating disorders want to quit their eating and exercise habits but cannot. In fact, some want to, some do not want to, and the same sufferers might want to at some times and not at others. Would that mean that they are only sometimes addicted? Second, this would mean that addiction is a state that comes and goes depending on what the person wants. The paradox here is that, for example, a heavy smoker who has no wish to quit is not addicted, whereas the heavy smoker who has a wish to quit, but cannot or finds it very hard to quit, is an addict. Surely, if a heavy smoker is addicted, s/he will be addicted whether or not s/he wants to quit.

Finally, one could argue that a person is addicted when they have a devotion or a habit that is *harmful*, and *cannot quit it* (could be impulsive or compulsive action according to the definitions given earlier), *whether or not they want to quit.* So, for example, the smoker is addicted *if s/he cannot quit* smoking, or finds it very hard to do so, even if s/he has no wish to do so. But this again makes any risky or harmful habit addictive. Under this definition, those who are exercise adverse, for example, and thus endanger their health by not taking enough exercise, and those who find it too hard to adhere to a healthy exercise program, would all be addicted. They have habits that are harmful and they find it very hard to break them. Also, there would be no difference between the devotion of the journalist and smoking, and sedentary life, and eating-disordered behavior. All of them are forms of risky devotion or harmful habits, under the definition, and difficult to give up. The weak meaning of addiction, even in the modified versions, even with these further specifications, is still too far-reaching and all-encompassing.

Because of these problems, if eating disorders and over-exercise are addictions, they must be addictions in the second and stronger sense of the word.

The strong meaning of addiction

In a strong sense, a person is addicted when s/he has a dependence on a substance that is psychoactive.[14] Thus, neurophysiological imbalances can, at least to a significant extent, explain his/her behavior (though concomitant explanations can also be relevant).

The idea of a common neurophysiological basis between eating disorders, excessive exercise and addiction is one that has appealed to many scientists. Finding a common biological basis between eating disorders, over-exercise and addiction would contribute to the specificity of the diagnosis of eating disorders and to the efficacy of treatment options.[15] Moreover, this correlation between eating disorders and addictions captures the experience of the sufferers.

Eating disorders and addiction

People afflicted by eating disorders share experiences[16] that are very similar to those of people afflicted by substance use dependence.[17] These similarities are summarized in Box 4.1.

Box 4.1 Similarities between eating disorders and addiction

- Eating-disordered behavior has overt adverse medical and social consequences, as does addiction. However, the sufferer in both cases persists with disordered eating and substance use.
- The person with bulimia feels an uncontrollable urge to eat, similar to the person with addiction who feels an urge to take the psycho-addictive substance.
- The person with eating disorders suffers loss of control, like the addicted; s/he constantly thinks of food, as the person with addiction constantly thinks of the substance of addiction.
- Food (or the substance) is utilized to cope with stressful situations and negative feelings.
- Both eating-disordered people and people suffering from addiction find it difficult to admit their problem and seek help, are ambivalent about treatment and have a high risk of relapsing.

There is also high co-morbidity (co-morbidity is the appearance of two conditions in the same person) between addiction and eating disorders, especially bulimia.[18] This means that many people with eating disorders, especially bulimia, also develop addiction to alcohol or other drugs. Also, for this reason, some scientists believe that there is a common basis for addiction and eating disorders. People with addiction and with eating disorders also share personality traits that, according to some, make them prone to addiction: for example, impulsiveness, affective instability, negative emotionality and sensitivity to rewards.[19]

Bulimia and dependence

The similarities outlined above in Box 4.1 concern more 'bulimic' rather than 'anorexic' restrictive behavior (fasting). The experience of bulimia is all about abusing food, losing control, feeling compelled to overeat and purge. In fact, some researchers argue that bulimia (and not anorexia) is a form of addiction.[20] There are a number of problems with this hypothesis.

First, if bulimia and not anorexia is a dependence, this means that anorexia and bulimia are different conditions. Indeed, some claim that anorexia and bulimia *are completely different syndromes*.[21] However, the vast majority of experts agree that they are the two sides of a common problem: excessive concern towards the body, terror of fat and weight.[22]

Second, it has been argued that bulimia generates dependence, because people with bulimia tend to prefer sweet carbohydrates during their binge (though sometimes the sufferer has no preference between sweet and salty carbohydrates),[23] and sweet foods, particularly simple carbohydrates (for example, glucose, fructose, galactose or disaccharides, which are found in foods like cakes, pastries and chocolate) generate dependence.[24] The fast assimilation of carbohydrates determines a sensitive increase of glycemia (the concentration of glucose in the blood). Glycemia, in turn, acts on the secretion of insulin. The more intense the secretion of insulin, the faster the decrease of glycemia ('hypoglycemia') in the blood. The result of this circuit of feedback is the new sensation of hunger. This circuit is called 'reactive hypoglycemia'.[25] One hypothesis is thus that bulimics become dependent on carbohydrates. However, where reactive hypoglycemia may *contribute* to explaining why bulimics tend to prefer some foods and why it is so difficult for them to give up bingeing, it does not explain how bingeing practices start.[26] More importantly, it is difficult to explain why bingeing is necessarily followed by purging, and also why all those with a 'sweet tooth' do not become bulimic.

Finally, the chemical dependence that occurs in addiction is characterized by *tolerance* and *withdrawal* reactions, which are absent in eating disorders.[27] The person with bulimia, for example, does not need increasingly higher doses of food during food orgies, and signs of withdrawal (like sweating, trembling, vomiting, diarrhea, fever) typical of withdrawal of psychoactive substances, are not found in bulimics who abstain from bingeing.[28]

Thus the hypothesis that bulimia is a form of addiction in the strong sense is at least incomplete.

Anorexia and dependence

According to some researchers, it is starvation and exercise rather than bulimia that can produce dependence. Marilyn Duker and Roger Slade write:

> Starvation [. . .] results in the secretion of adrenalin [. . .] [Adrenaline] acts on the brain and causes it to secrete endokinins. These are chemicals closely related to morphine and have similar tranquillising and euphoric effects. At the same time, metabolites (such as ketones) which are produced by the breaking down or metabolism of fat also act on the brain and can create an odd and lightheaded experience.
>
> [M]orphine-like substances (endorphins) [. . .] are also produced by [. . .] vigorous exercise. [. . .] This is how an individual can come to derive a particular pleasure [. . .] from strenuous exercise. [. . .] It is thus that anorexic illness can be viewed as an addiction to food/body control. Sufferers occasionally refer to themselves as 'starvation junkies' or as needing their 'exercise fix'.[29]

The question is whether the amount of neurotransmitters produced by starvation and exercise explains the harm that anorexics and over-exercisers inflict upon themselves. Even taking into consideration the pleasure of lightheaded experiences that can be provided by starvation and exercise, extreme diet and over-training are also sources of physical and psychological pain. We have seen in previous chapters the effects of starvation and abnormal nutrition: the person with an eating disorder suffers a great deal of physical stress and pain. Over-training can also be unpleasant and painful (see Box 4.2).

According to von Ranson and Cassin, who have performed a comprehensive overview of research on addiction and starvation, there is limited evidence that

---

Box 4.2 Common effects of over-training:[30]

- tiredness;
- lethargy;
- insomnia;
- irritability;
- depression;
- flatulence;
- diarrhea;
- clamminess;
- night sweats;
- anxiety;
- frequent urination;
- loss of appetite;
- loss of interest in sex;
- abnormal heart rate patterns.

starvation and exercise produce enough endorphins to explain the excesses of anorexia and over-exercise.[31] Indeed, if this was the case, the vast majority of those who engage in regular diet and physical activity would develop anorexia, which is certainly false.

Research, thus, offers no conclusive proof that diet or exercise produce physical dependence, which might justify the extreme behaviors characterizing anorexia and bulimia nervosa.

Exercise and dependence

We saw at the beginning of this chapter, as well as in the Introduction to this book, that many see exercise as a potential threat to health and a potentially addictive activity. Sometimes it can be seen as a precursor to eating disorders or even a cause of them. But according to research, exercise *can prevent* the onset of eating disorders, as it improves body satisfaction. Some have suggested that the reason why more exercisers than inactive people have eating disorders might be that people with a predisposition to eating disorders tend to gravitate towards various forms of exercise.[32] Certainly there is an association between eating disorders and excessive exercise.[33] However, it is not clear which is the cause and which the result. Eisler and Le Grange[34] have proposed that eating disorders and over-exercise might interact in four ways:

1   Over-exercise and eating disorders are distinct phenomena, which look similar because in both cases the person engages in exercise of a duration and intensity that is self-harming.
2   Over-exercise and eating disorders are overlapping phenomena, each increasing the chance of developing the other.
3   Over-exercise and eating disorders are ramifications of an underlying psycho-logical problem, most likely obsessive-compulsive personality traits, and this is why they often occur together.
4   Over-exercise and eating disorders are variants of each other, and environ-mental factors such as familial, sexual and cultural factors account for why one person expresses a basic vulnerability through abnormal eating and another one through excessive exercise patterns.

In any event, the idea that exercise initiates eating disorders or can be addictive is to an important extent speculative and incomplete. There is neither something inherently dangerous in exercise and healthy nutrition, nor something so irresistibly pleasurable to compel people to exceed in frequency and intensity of exercise or starvation. The person who starves does so despite all secondary symptoms suffered. Likewise, the person who continues to over-exercise does so despite all secondary symptoms suffered. Of course, the alcohol and heroin addict also keeps taking the substance despite the ill effects on his health and social life: but whereas tolerance, withdrawal and other neurophysiological processes contribute to explain why the urge to keep taking the psychoactive substances is strong enough to overcome those ill effects, these further explanations are absent in eating disorders and exercise. Moreover, whereas taking excessive alcohol and other drugs is inherently pleasurable – these drugs produce pleasure – starvation and fatigue are not. Thus a more complete hypothesis relating to the etiology of eating disorders and over-exercise must be formulated.

Elsewhere[35] I have argued that the key for understanding eating disorders, and here I include over-exercise, is, in an important way, to be found in the pleasure obtained through the exercise of self-control and mastery over the body. The root of eating disorders and other abuses of the body is thus not primarily physiological, as the addiction model would suggest, but primarily moral. I shall develop this hypothesis in Chapter 5. Briefly, it seems that the one advantage that might make eating-disordered behavior understandable is the sense of empowerment and control over the body that the eating-disordered person gets out of her rigid eating and exercise routines. This sense of empowerment, however, can only be explained in light of a moral background, which disparages physicality and places high value on the spirit, on the strength of the will, and which sees control of the body, and all that comes with it, as a worthwhile goal.

This moral background, as I shall further clarify in Chapter 5, has appealed to many religions, but is more ancient than Christianity itself, and is based on a particular metaphysics of the human person. According to this metaphysics, human beings are divided into a body and a mind, which are different substances. The mind, or spirit, is valuable, and the body is corruptible, and thus potentially dangerous. It therefore needs to be mastered and controlled. Eating disorders are difficult to understand without reference to these metaphysical and moral ideas, according to which self-control and self-mastery are measures of strength of character.[36]

From this perspective, we can understand why overeating generates feelings of guilt and shame. Corpulence – literally abundance of body – is a sign of weakness; fat is repugnant in that it expresses the person's 'incapacity' to master the body. Desire to eat or longing for food, which is an inherently healthy drive for life, is experienced as 'bad' and shameful. I have discussed the link between metaphysical dualism and hunger elsewhere[37] and my observations cannot be repeated fully here. I will articulate this discourse on the metaphysics of individuals and hunger a little more in Chapter 5. Here, however, it is important to note that I am not suggesting that the metaphysical dualism 'causes' eating disorders. Metaphysical dualism, as I shall argue in Chapter 5, is much older than modern eating disorders. Moreover, as, again, I have discussed elsewhere,[38] in other cultural groups that adopt strong dualistic views of the individual, eating disorders are not as widespread as in Western countries. I am instead suggesting that some aspects of eating disorders can be understood with reference to a certain conception of the individual. If we want to understand eating disorders, we need to understand how people normally think about themselves, about personhood. Metaphysical dualism thus does not necessarily cause eating disorders, but contributes to explaining them.

From this perspective we can also understand better why dieting is experienced by the sufferer as empowering. The apparent paradox that lurks in anorexia – the smaller and weaker one becomes through dieting, the stronger and more invulnerable s/he feels – becomes more intelligible in the light of a moral background.

I will not go any further here with my arguments about the moral basis of eating disorders. What matters to us now is that the hypothesis that eating disorders and over-exercise are forms of addiction is in an important sense incomplete. The implications of this for fitness and sports practice will be discussed in the Conclusions to this chapter and in Chapter 9.

Let us get back to Gus Goose.

Is goodness bad?

The claim that eating disorders and exercise are addictions can be misleading in another sense: it can lead people to demonize exercise, and to believe that exercise and diet can take over our life, and that we, without any willful control, can spiral into a dangerous pattern of behavior that can put our health and life at risk. The facts contradict any such suggestion. There is nothing inherently dangerous or addictive in exercise and good nutrition. The expected benefits of moderate and vigorous physical activity vastly outweigh potential risks.[39]

It is instead a sedentary life that is the biggest killer of our times.[40] Estimates by the World Health Organization (WHO) and the Organisation for Economic Cooperation and Development (OECD), as well as an analysis of European countries,[41] show that in industrialized countries the leading causes of death are chronic illnesses such as heart disease, cancer and depression. The WHO has drawn up a list of the major chronic conditions and causes of death in industrialized countries (see Box 4.3).[42]

Box 4.3 Major causes of death in industrialized countries:

- cardiovascular diseases;
- hypertension;
- stroke;
- diabetes;
- cancer;
- chronic obstructive pulmonary disease;
- musculoskeletal conditions (such as arthritis and osteoporosis);
- mental health conditions (mostly dementia and depression);
- blindness and visual impairment.[43]

These conditions are strongly related to obesity, and obesity is strongly related to lack of physical activity (see Chapter 6 for a definition of obesity). Research has shown that over the last century being overweight "has become increasingly common despite an overall decrease in the average daily caloric intake of the population [. . .] physical activity is inversely related to body weight, body composition, and waist-to-hip ratio".[44] Children who watch television five hours a day are 8.3 times more obese than children who watch television two hours a day.[45] The interpretation of these data is not straightforward: it is unclear whether lack of physical activity is a consequence of being overweight or a cause of it. Moreover, research has not always taken into account that the obese person, in order to move, expends more energy than the thin person, because the weight s/he has to carry is more, therefore lower physical activity might not mean necessarily lower energy expenditure. What remains uncontroversial, however, is that physical inactivity is associated with obesity and is a leading cause of death, and physical activity is not. The relationship between cardiovascular disease, hypertension, stroke, diabetes, musculoskeletal conditions and exercise has been established.[46] Physical activity (in different forms depending on people's age, level of fitness and eventual medical conditions) prevents the occurrence of these conditions.[47]

People in Western countries are increasingly more sedentary. Yet, like Gus Goose, nobody has ever been described as 'repose-addicted', and the psychiatric category of 'Physical Inactivity' has yet to be included in diagnostic manuals.

The point I want to make is not that the obese and sedentary person should be regarded as mentally ill, but that the terms 'addicted' and 'dependent' are far from neutral or purely descriptive. They are value-laden. They refer to something bad and undesirable. These terms traditionally denote the abuse of psychoactive drugs.[48] So, when people characterize something as 'addictive', that something is associated with an undesirable state, the state of being dependent upon a dangerous thing, of being at high risk of serious harm or death, and of being out of control. Saying that exercise and diet may be addictive implies that exercise and diet are *potentially harmful or lethal*. But there is nothing intrinsically harmful in exercise and physical activity, and a sedentary life is overall much more risky than an active life. Some people may eat too much and thus harm themselves, but food is not dangerous, and the same goes for exercise. The abuses of eating and exercise that we find in some individuals should not lead us to blame or demonize physical activity, sports and healthy nutrition.

Conclusions

It is sometimes argued that eating disorders and exercise are forms of addiction. Scientific evidence, however, seems to discredit this hypothesis. In fact, if it were true that dieting and exercise were so pleasurable, the epidemics of obesity, sedentary life and related diseases would be difficult to explain. This chapter has shown that eating disorders and exercise are not forms of addictions, and are not addictive.

This finding is important to the purposes of this book. People with eating disorders are, to an important extent, in charge of their choices. This does not mean that it is easy for them to change their behavior or for fitness professionals to help them. It means that people affected have the resources to improve the quality of their life. For fitness professionals, this means that the eating disorders sufferer and the over-exerciser are not 'ill' clients, who must be handed over to medics. Eating disorders have serious adverse medical effects. For this reason, it is appropriate that the physical sequels associated with malnutrition are duly managed by medically qualified professionals. It is also of course appropriate that fitness leaders take appropriate measures to avoid preventable harm or injury to the physically vulnerable exerciser (see Chapter 8). However, it is not clear that eating-disordered exercisers are truly 'slaves' of their habits. There is reason to consider them as in principle capable of negotiating a healthier exercise pattern, and it should not be assumed that it must be in the client's best interests to be excluded from recreational activities.

The fact that eating-disordered exercisers are to an important extent the 'deciders' also raises a sensitive ethical issue: are interferences with their freedom to exercise justified? Why should it be all right to exclude an eating-disordered exerciser from the gym for their own good, when other people are free to do risky sports, such as free climbing, parachuting and racing? Why should it be legitimate to force some not to exercise, and not legitimate to force the sedentary person to exercise? And what does this imply for fitness professionals? Does it mean that because eating-disordered exercisers have, in principle, free will, they must supervise them?

The next chapter will discuss another widespread argument: that eating disorders are a result of social pressure, often expressed through the media, to be thin.

Key points

- The Addiction Model regards addiction as essential to the etiology of eating disorders and over-exercise.
- In the weak sense, addiction means being given to a habit or pursuit; or devotion.
- In the strong sense, addiction means being addicted to a drug; an urge and need to continue taking a drug as a result of taking it in the past.
- If eating disorders and over-exercise are addictions in the weak sense, either they are normalized, or many normal activities are pathologized. Therefore the weak sense of addiction cannot be applied to eating disorders and over-exercise.
- Eating disorders and over-exercise are not forms of addiction in the strong sense, because the neurophysiological processes related to them do not explain the severity of the behavior.
- The Addiction Model is incomplete.
- Exercise and diet should not be demonized purely because they are abused by some people.
- There are insufficient grounds to regard exercise and diet as intrinsically addictive.

5 Media and eating disorders

Introduction

In ordinary discourse, but also in medical settings (see the next section of this chapter), eating disorders are often related to the pressure from the media to be thin. The argument is that the wide use of super-thin models inculcates the wrong ideals of beauty and wrong ideas of what it is to be normal. In the attempt to mimic the idolized models of beauty, young women – so the argument goes – begin to diet, and then are unable to stop. Many people believe that there is something pernicious in the idea that, in order to be beautiful, a person needs to be very thin and/or strong, toned, smooth and statuesque. This chapter assesses the soundness of these claims. It addresses the question of whether the pressure operated by the media can be responsible for eating disorders, and, if so, in what way.

This is the last chapter devoted to the causes of eating disorders. Understanding eating disorders, and what can cause or trigger them, is essential to the framing of management plans. If eating disorders are the result of media pressure, the fitness industry could be called on to correct negative models of beauty and normality, and to propose alternative and more functional models of reference for the public. Whereas it is important that, within the fitness industry, misleading models of beauty or health be corrected, and whereas it might be beneficial to the general public if the media reduced the broadcasting of emaciated models, I argue that the media are not responsible for eating disorders and over-exercise. Eating disorders and over-exercise stem from a fear of fat and weakness, which cannot purely be explained in terms of social pressure to be thin. The 'media' or 'fashion' arguments essentially say that people prefer thinness because society prefers thinness – a circular argument. The use of thin models in the media, I argue, has a moral background. This moral background needs to be acknowledged, in order to understand eating disorders.

Princess Diana, in 1993, gave a speech on eating disorders. This testimonial represents an important statement on the responsibility that society has towards young people and towards sufferers of eating disorders in particular.

Ladies and Gentlemen.
I have it, on very good authority, that the quest for perfection our society demands can leave the individual gasping for breath at every turn.

This pressure inevitably extends into the way we look. And of course, many would like to believe that eating disorders are merely an expression of female vanity – not being able to get into a size ten dress and the consequent frustrations!

From the beginning of time the human race has had a deep and powerful relationship with food – if you eat you live, if you don't you die. Eating food has always been about survival, but also about caring for and nurturing the ones we love. However, with the added stresses of modern life, it has now become an expression of how we feel about ourselves and how we want others to feel about us.

Eating disorders, whether it be anorexia or bulimia, show how an individual can turn the nourishment of the body into a painful attack on themselves and they have at their core a far deeper problem than mere vanity. And sadly, eating disorders are on the increase at a disturbing rate, affecting a growing number of men and women and a growing number of children.

Our knowledge of eating disorders is still in its infancy. But it seems, from those I have spoken to that the seeds of this disease may lie in childhood and the self-doubts and uncertainties that accompany adolescence. From early childhood many had felt they were expected to be perfect, but didn't feel they had the right to express their true feelings to those around them – feelings of guilt, of self-revulsion and low personal esteem. Creating in them a compulsion to 'dissolve like a Disprin' and disappear.

The illness they developed became their 'shameful friend'. By focusing their energies on controlling their bodies, they had found a 'refuge' from having to face the more painful issues at the center of their lives. A way of 'coping', albeit destructively and pointlessly, but a way of coping with a situation they were finding unbearable. An 'expression' of how they felt about themselves and the life they were living.

On a recent visit to the Great Ormond Street Hospital for Sick Children I met some young people who were suffering from eating disorders. With the help of some very dedicated staff, they and their parents were bravely learning to face together the deeper problems which had been expressed through their disease.

With time and patience and a considerable amount of specialist support, many of these young people will get well. They and their families will learn to become whole again. Sadly, for others it will all be too late. Yes, people are dying through eating disorders.

Yet all of us can help prevent the seeds of this disease developing. As parents, teachers, family and friends, we have an obligation to care for our children. To encourage and guide, to nourish and nurture and to listen with love to their needs, in ways which clearly show our children that we value them. They in their turn will then learn how to value themselves.

For those already suffering from eating disorders, how can we reach them earlier, before it's too late?

Here in Britain organizations such as 'The Eating Disorders Association' are currently being swamped with enquiries and requests for support and advice, so overwhelming is the need for help. Yet with greater awareness and more information these people, who are locked into a spiral of secret despair, can be reached before the disease takes over their lives. The longer it is before help reaches them, the greater the demand on limited resources and the less likely it is they will fully recover.

I am certain the ultimate solution lies within the individual. But with the help and patient nurturing given by you the professionals, family and friends, people suffering from eating disorders can find a better way of coping with their lives. By learning to deal with their problems directly in a safe and supportive environment.

Over the next three days, this International Conference has the opportunity to explore further the causes of eating disorders and to find new avenues of help for those suffering from this 'incapacitating disease'.

I look forward to hearing about your progress and hope you are able to find the most 'beneficial' way of giving back to these people their self-esteem. To show them how to overcome their difficulties and redirect their energies towards a healthier, happier life.

(Speech given by Diana, Princess of Wales, at the opening ceremony of an international conference on eating disorders in London, 27 April 1993)[1]

Princess Diana has been loved worldwide for her generosity. This speech was one of the many good things she gave to the community. She also declared publicly that she herself suffered from eating disorders, and thus her speech becomes even more meaningful, because it is not just the Princess's recognition of the drama suffered by eating-disordered people and their families, but also the moving testimonial of a sufferer, a person like many, with ordinary weaknesses and sorrows.

I had bulimia for a number of years. And that's like a secret disease. You inflict it upon yourself because your self-esteem is at a low ebb, and you don't think you're worthy or valuable. You fill your stomach up four or five times a day – some do it more – and it gives you a feeling of comfort. It's like having a pair of arms around you, but it's temporary. Then you're disgusted at the bloatedness of your stomach, and then you bring it all up again. And it's a repetitive pattern, which is very destructive to yourself.[2]

Princess Diana's candid views were broadcast to the world, and also helped dismantle some misleading stereotypes on eating disorders. Eating disorders might be triggered or exacerbated by the pressure to be thin operated, as she suggested, by 'society'. But, as Princess Diana recognized, there is more to eating disorders than a vain pursuit of beauty, or a need to reach the proportions of models portrayed in the media.

Eating disorders and over-training in common discourse

In common discourse, eating disorders are often related to the pressure to be thin, to the predominant use of very thin models on catwalks and in the media.

The medical profession has also raised worries that the nearly exclusive use of extremely thin models could contribute to the spread of eating disorders. In the UK, the British Medical Association (BMA) expressed its concern about the use of very thin models within the fashion industry, and suggested that the media can "play a significant role in the aetiology of eating disorders", that it can "'trigger' the illness in vulnerable individuals, by suggesting that being 'thin' means being successful".[3] Similarly, the Institute of Psychiatry in London has also been persuaded by the argument that the media may have some responsibility for eating disorders.[4]

The publication of these reports has elicited opposing reactions. Media representatives did not appreciate the reports, finding them inappropriate and misleading. Publishers of women's magazines said that the blame should not fall on them: the reason they use super-thin models is that people tend to buy magazines where skeleton-like bodies are pictured.[5] Is it their fault if this is what the public want to see? I will return to this response in later sections.

Many instead applauded the reports. The first government to take this on board was the regional government of Madrid which, in 2006, banned overly thin models at top-level fashion shows in Madrid. The CCN reported:

> Madrid's regional government, which sponsors the show and imposed restrictions [. . .] said the fashion industry had a responsibility to portray healthy body images. [. . .] Organizers say they want to project an image of beauty and health, rather than a waif-like, or heroin chic look.[6]

The major of Milan, Letizia Moratti, wanted to impose a similar ban. Girls with a Body Mass Index of below 18, which is classified as medically underweight, would be banned from shows in Milan. New York's Elite modeling agency claimed that a similar ban was discriminatory against many models, and would curtail the freedom and creativity of designers. The UK's Marks & Spencer also decided not to follow the Spanish move. The head of Marks & Spencer declared it was up to the designers to decide the size of their models.[7] *Dove*, the producer of deodorants and soap, instead, introduced 'normal size models' in its ads, encouraging the use of models who had Body Mass Indexes between 18 and 25, which is in accordance with United Nations guidelines for health.[8]

The UK in general did not follow the Spanish example: it was believed that the decision on the models' size should rest with the designer, and it would not be a matter for legislation.

Some comments

In this debate, the arguments of the fashion industry are ill-founded. When images are believed to cause psychological or social damage, the state intervenes to restrict, in various ways, their projection. This is why the broadcasting of violent films and other images is regulated. The limits that the law sets to the broadcasting of strong language and films is not thought to impinge negatively on the creativity and freedom of film directors and the public at large. If it does, that infringement is justified in terms of people's health and safety. It is not clear why this general rule should not apply to fashion. Moreover, where it might be true that excluding very thin models from a modeling career is discriminatory, it is equally discriminatory to informally exclude those with larger bodies from a modeling career.

Thus, in principle, there is scope for legislation or for governmental action in regulating the use of models in the fashion industry and the media. Equal opportunities should be given to all people, not only to the extremely thin. The law should protect workers from having to accept unhealthy working conditions (in this case, unhealthy diets) in order to work. And finally, if it were true that the broadcasting of images of super-thin models had negative impact on public health, this would constitute a good ground to impose limits.

But *is it true that the media contribute to the onset of eating disorders and of other abuses of the body?* The answer, I argue, is not straightforward. In order to understand the complex relation between aesthetic norms (what we should look like in order to be 'beautiful' or acceptable) and the role of the media and fashion (which might seem to dictate or create and sustain these rules) it might be useful to expand our perspective and consider how women's bodies are used in other contexts, such as the arts.

Fashion and the arts have several points in common. They both create beauty and represent beauty; they both dictate norms of beauty; and they both use human bodies as models. Vanessa Beecroft has revealed these points of contact between fashion and the arts in her own art. I now turn to her artworks, as they can illustrate some of the complexities that lie behind eating disorders.

Vanessa Beecroft: The woman in the public eye

Vanessa Beecroft is a contemporary artist who has become famous for her displays of female models. The performances exhibit models standing for hours. The models are often naked. Sometimes all they wear is one accessory, whether high heels, tights or a wig.

Models remain silent and as immobile as possible, in the position chosen by the artist. Performances last a long time, sometimes eleven hours, leaving the models bored and exhausted.[9] The models by contract have to obey the artist's orders, which are basic and rudimentary, such as 'do not move' and 'do not talk'. During the exhibition the models are allowed to stretch or move only if necessary, and towards the end of the performance they are allowed to lie or sit on the floor.[10] Beecroft's models are reduced to uniformity by means of their own nudity, and by the conformity of the one accessory that is imposed on them all. Each woman suppresses her identity by contract with the artist. Nudity is not an expression of sensuality or sexuality, but of a cruel homogeneity. Figures 5.1 to 5.5 show some of Beecroft's performances. Women are rendered as identical as possible, and this homogeneity of silence and immobility which swallows individuality and reduces every single woman to *an identical other*, creates a disquieting atmosphere.

Her works and the philosophy behind them express some profound issues concerning the place of the woman in society, and can tell us something about the complex relation between the media, beauty and eating disorders.[11]

The way Beecroft uses 'the girls', as she calls them, in her exhibitions is the artistic replication of the way in which women seem to be used in the fashion industry and sometimes in the media. In order to be adequate to their role, the models must be homogenized, and must conform to accepted standards of beauty. Their subjectivity is filed away.

Of course, in the arts, standardization of beauty has been common practice, at least since the Renaissance. The Renaissance codified geometric proportions of beauty, which had to be used in the representation of human beings.[12] In this sense the arts are somehow intrinsically 'eugenic'. But as the eugenic project of

Figure 5.1 Vanessa Beecroft, VB48.721.dr , VB 48 Palazzo Ducale, Genoa, 2001
©2009 Vanessa Beecroft. Courtesy of Galleria Lia Rumma & Massimo Minini

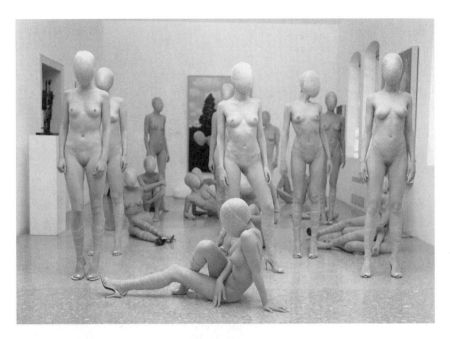

Figure 5.2 Vanessa Beecroft, VB45.107.dr, VB 45 Kunsthalle Wien, Vienna
©2009 Vanessa Beecroft. Courtesy of Galleria Lia Rumma & Massimo Minini

Figure 5.3 Vanessa Beecroft, VB47.738.dr, VB 47 Peggy Guggenheim Collection, Venice
©2009 Vanessa Beecroft. Courtesy of Galleria Lia Rumma & Massimo Minini

Figure 5.4 Vanessa Beecroft,
VB52.02.nt, VB 52 Castello
Di Rivoli, Turin

©2009 Vanessa Beecroft.
Courtesy of Galleria Lia Rumma
& Massimo Minini

Figure 5.5 Vanessa Beecroft,
VB52.04.nt, VB 52 Castello
Di Rivoli, Turin

©2009 Vanessa Beecroft.
Courtesy of Galleria Lia Rumma
& Massimo Minini

the arts is limited to canvas or marble, this has raised no special ethical concerns. When, instead, the arts use humans (and not just canvas or marble) as *living subjects*, as *living models*, the atmosphere created becomes "sinister and foreboding".[13] Both Beecroft and the fashion industry do so: they use living beings as representations of a certain conception of beauty and acceptability.

Vanessa Beecroft has had enormous success. She has probably caught human imagination and feelings relating to something that matters to many people: the exhibition of the self and annihilation of identity in the public eye. Beecroft has given a disquieting representation of the fact that the woman who falls in the public eye is, and must be, stereotypical. She is a living piece of art, and as such she must conform to stated standards of acceptability – standards whose mere existence dictates the cessation of individual life. The voyeurism that people demonstrate when they watch with some consternation the catwalks and magazines populated by anomalously emaciated models is replicated in the galleries, where visitors gaze with perturbation at these women whose mere existence spreads an interesting malaise.

Women in the public eye, thus, cannot have their own distinguishable and different body, their different and differentiated identity. Women must be homogenized, de-carnalized, deprived of identity. Not surprisingly, therefore, campaigns against the use of skinny models have largely been ineffective. The recommendations of the Spanish government have been rejected by most other governments. Catwalks, as well as the media more generally, continue to be populated by superthin models.[14]

The media thus simply *respond* to a widespread preference for certain types of bodies which symbolize not just an aesthetic preference, but rather a set of values and beliefs relating to what people should be. Seemingly *aesthetic* preferences expressed in the media are not just aesthetic: they are value-laden. This means that the causation link between media and body abuses is twofold: people have a strong preference for thin and strong bodies deprived of fat, and this is why the media utilize them. They sell what the public wants to buy. On the other hand, media can reinforce public preference, normalize it, and give further strength to the values that are shared and that lie behind these preferences. In this sense, the media can in some way trigger the onset of eating disorders, as we are now going to see.

The impact of the media on eating disorders and exercise

In research involving people with bulimia, participants have stated that portrayal of women in the media is one of the factors that are likely to contribute to the onset of bulimia nervosa.[15] The Eating Disorders Association has also argued that the media can trigger the onset of eating disorders and other abuses of the body.[16]

Mervat Nasser noted that

> the media is seen to present images of thinness through role models or through images that imply social desirability [. . .] women were found to use

media images as a reference source in evaluating their own body image and also their sense of acceptance or approval.[17]

Psychology studies show that congruence with the accepted models of physical adequacy and beauty appears to influence self-esteem. Women's and, to a lesser extent, men's self-esteem is susceptible to manipulation operated by what is portrayed by the media.

> Seeing photographs of thin fashion models led to a reduction in women's general sense of self-esteem. [. . .] Other research confirmed this effect, particularly on adolescent groups. [. . .] Women also appear to be more easily manipulated by media images than men although there is an indication that men's self-esteem too is influenced by what is portrayed in the media.[18]

According to some research, body dissatisfaction generally – not only eating disorders and over-exercise – can result from this type of influence, and this explains why increasing numbers of people, especially women, resort to surgery to modify their physical appearance. Ogden has labeled this as "the surgical age".[19]

Certainly, it can be argued that the male and female prototype body – devoid of body fat and remarkably toned – which is spread by the media is unrealistic and standardized. It leaves no room for differences in somatotypes, shapes, skin types and muscle types. Each of us has his/her own genetic make-up, and this cannot be altered by diet or exercise. Figure 5.6 gives examples of the main somatotypes.

An endomorph, however much s/he diets, will never reach the proportions of Kate Moss. An ectomorph, however much s/he trains, will never have huge muscles of the Rambo type.

Moreover, messages about 'healthy weight' often connect 'health' with a narrow range of body shapes and weight. According to research, paradoxically, 'health messages' often provide a rationale for binge eating and compensatory behavior, which people with eating disorders often learn from magazines and the media: many 'health farms' advertise methods for 'cleansing' the organism and regulating energy expenditure as healthy practices.[20] I shall get back to health farms at the end of this chapter.

Whereas many report that there is an evident link between eating disorders and media images, research on the history of beauty allows us to look beyond this simple link, and to understand the values that lie behind what people *admire*.

Muscles and masculinity

Why a man, in order to appear 'attractive', has to display a certain degree of muscle mass is relatively easy to understand. Physical strength is congruent with gender stereotypes of masculinity crystallized in Western culture. Manly force corporealizes the ideals of male protectiveness, strength and courage. In popular

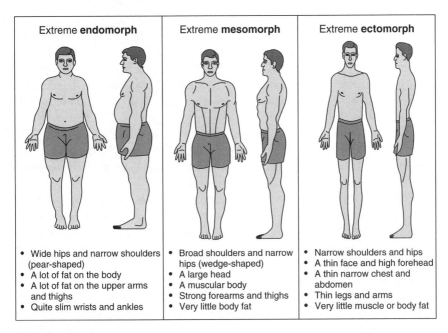

Extreme **endomorph**	Extreme **mesomorph**	Extreme **ectomorph**
• Wide hips and narrow shoulders (pear-shaped) • A lot of fat on the body • A lot of fat on the upper arms and thighs • Quite slim wrists and ankles	• Broad shoulders and narrow hips (wedge-shaped) • A large head • A muscular body • Strong forearms and thighs • Very little body fat	• Narrow shoulders and hips • A thin face and high forehead • A thin narrow chest and abdomen • Thin legs and arms • Very little muscle or body fat

Figure 5.6 Somatotypes

Available at: http://img131.imageshack.us/img131/9071/somatotypes10sitp8.jpg (accessed 4 December 2009).

culture, physical strength is also often associated with moral integrity. Heroes often combine physical strength and moral rectitude.[21]

The link between muscular strength, masculinity and moral character is therefore easy to understand, in the light of stereotypes that are internalized in our culture. The link between femininity, thinness and muscular strength is, instead, more difficult to capture.

Femininity, thinness and muscularity

Much research has addressed the body size and shape of media images of women. Historical analyses of images of women have reported that the preferred woman's body has become consistently smaller over the past century than it had ever been before this time. [. . .] For example, from the Middle Ages the rounder 'reproductive figure' was considered attractive and plumpness was erotic and fashionable. In line with this, the women painted by Rubens in the 1600s had full rounded hips and breasts, and in the 1800s Courbet painted women who could be considered fat by today's standards.[22]

The 1900s, in particular, have witnessed a shift in tastes for women: from the Sophia Loren type to the Twiggy and Kate Moss type, and, more recently, the super-fit, big-shoulders–narrow-hips type. This shift is more difficult to understand. The shape of a woman has clearly been masculinized during the twentieth century. Female sex characteristics, such as breasts, hips or thighs, which corporealize femininity and sensuality, leave little room for muscular and androgynous bodies.

It is not surprising that what is regarded as sensual or beautiful changes over time. Our tastes for clothes and hats and cars change. It appears obvious that our tastes in the appearances of women and men also change. But why do they change? Studies on the history of thinness show that thinness has been valued at times when women were expected to demonstrate their intellectual skills. The more women aspire to 'male' positions, the more pursue a cylindrical or tubular (androgynous) body.[23] Thinness, thus, is not *intrinsically* beautiful. It is valued for what it symbolizes in a particular social context.

Where women have fought for equal opportunities, they might find it difficult *to be represented differently*. The woman has to show her strength to be respected as an 'equal'. A masculinized body is thus the symbol of *gained equality*. Research involving heterosexual men and women and homosexual men shows that weight and body shape concerns are greater for heterosexual women and homosexual men than they are for homosexual women and heterosexual men. The social role in which a person identifies him/herself influences the way s/he experiences his/her body and the way s/he wants to look and wants to be perceived by others.[24]

Let us now go back to the magazine publishers, and their response to the BMA

Figure 5.7 Sophia Loren

report. The BMA pointed out that magazines and the media could play a crucial role in the spread of eating disorders, and the publishers of women's magazines complained that it would be unfair to blame them for reproducing what people want to see. This response is important because it highlights the fact that people do not simply like thin and muscular bodies *because magazines are replete with these models*. It is (or, it is also) the other way round – magazines are full of thin and muscular models *because this is what people like*. It is not just that *what we see, we want*. Instead, in an important sense, *what we want, we see*. The question, therefore, is: what is attractive about thinness and muscularity?

Part of the answer has already been given: for many, thinness symbolizes a new place in society, new roles, novel ways of being perceived by others, and research on this has consistently upheld this hypothesis over the last 40 years.[25] But there is more than this to eating disorders and over-exercise. Changes in social roles might explain modifications in aesthetic preferences – how people see themselves, and how they want to be perceived by others. These changes might also explain why people tend to conform to ideals of beauty that are value-laden: more cylindrical bodies, for example, as opposed to rounded bodies. However, social changes still do not explain the excesses that we find in eating disorders. Misery and death cannot be purely explained in terms of modification of social roles.

Morality and abuses of the body

The excesses found in eating disorders and other abuses of the body can be best made sense of in light of moral values. We have already seen this in Chapter 4, and now I will discuss it in greater detail.

Thinking that eating disorders and over-exercise are a result of vanity or of veneration of the body is a gross misanalysis of the problem, as Princess Diana also recognized. These phenomena result, on the contrary, from denigration of the body. Denigration of the human body has a long-standing history.

Anna Krugovoy Silver writes:

> Historically, the body has been [. . .] denigrated and reviled as inferior and needing to be disciplined, punished, and ultimately transcended. In one classic, foundational statement of such a body/mind split, Socrates argues, in Plato' s "Phaedo", that "as long as we have a body and our soul is fused with such an evil we shall never adequately attain . . . the truth. The body . . . fills us with wants, desires, fears, all sorts of illusions and much nonsense, so that . . . if we are ever to have pure knowledge, we must escape from the body and observe matters in themselves with the soul by itself". Socrates' language associates the body with corruption, infection, "contamination", and "folly" that keeps a human being from the knowledge that can come only through the reasoning of the soul.[26]

Brief historical background

Human beings have been considered as characterized by two substances, body (matter) and soul (immaterial spirit), at least since Orphism.[27] Orpheus was a (probably legendary) Greek poet. According to Orphism, the human being is composed of soul and body. The soul is a *daemon* (δαίμων), a divine principle that fell into the body because of an original sin. The soul is immortal and reincarnates in different bodies, until the rituals and practices of the 'Orphic life' put an end to the cycle of reincarnations (*metempsychosis*) and the soul is set free. The body is the animal side, whereas the soul (or mind, or spirit, or rationality according to the language of choice) is the most distinctive side of human beings, what is inherently ours, and what distinguishes us from beasts.

This schema of thought had an irreversible effect on original Greek naturalism: for the first time the human being was presented as composed of two sides *in contrast* with each other, and physical impulses were presented *as something that needed to be transcended* (to be risen above). Orphism had a major impact on Greek culture, including philosophy and science.

Plato, for example, writes: "we shall be closest to knowledge if we refrain as much as possible from association with the body or join with it more than we must, if we are not infected with its nature but purify ourselves from it until the god himself frees us."[28]

Morality is on the side of the mind (or soul or spirit), and the body is inferior and corruptible.[29] This conception of the human being has appeared in Western philosophy and religious thought in all eras.[30] It has been accepted within Christianity, in the patristic doctrines, Scholastic philosophy, Humanism, during the Renaissance, and in contemporary philosophy.

The idea that body and mind are two different entities is so widely accepted in Western culture that the philosopher Gilbert Ryle has called it the 'Official Doctrine'.

> The official doctrine, which hails chiefly from Descartes, is something like this. With the doubtful exceptions of idiots and infants in arms every human being has both a body and a mind. [. . .] In consciousness, self-consciousness and introspection he [the man] is directly and authentically apprised of the present states and operations of his mind. [. . .] It is customary to express this bifurcation of his two lives and his two worlds. [. . .] It is assumed that there are two different kinds of existence or status. What exists or happens may have the status of physical existence, or it may have the status of mental existence.[31]

This metaphysics, although debated in philosophy, is still influential, and many people, including contemporary ethicists, have taken it for granted. That we *have* rationality or a mind seems so self-evident that it does not require any justification. If we have a mind, we must *also have a body* – as something ontologically (that is, in its substance) different from the mind.

This conception of the human being, which appears self-evident to many is not only not self-evident, but is also *not neutral*. It is covered in value. In fact, in this partition, the physical side is clearly the less valuable, the corruptible one, and the one that needs to be controlled and suppressed. Viewing physical life as something that can and should be controlled and submitted to the will may have an obvious impact on the way people perceive their physical impulses, including *hunger*.

This does not mean that if you are a Christian or have read Plato and Scholastic Philosophy you will become anorexic! Likewise, I am not suggesting that eating-disordered people are 'moral' and those who do not have eating disorders are 'immoral'. Nor am I suggesting that disordered eating practices are 'moral' practices. People with eating disorders should neither be applauded nor blamed for their condition, and a judgment of their condition is not what I want to convey. What I am suggesting is that the doctrines of many philosophers and theologians express values and ideas that are crystallized in Western culture, in the way many of us see the world. These values help us understand eating disorders and other abuses of the body. To say that people torture their body out of vanity or out of excessive love for their body has little credibility. Eating disorders and over-training are a form of torture, and people inflict it on themselves because, at some level, they devalue and denigrate their body, and have absorbed the belief that the body must be controlled and mastered.

Exercise and morality

The word 'exercise' comes from the Greek *askesis*, which means *physical* exercise or practice. *Askesis* mainly referred to the physical training of athletes.[32]

The early Christians translated *askesis* (physical exercise) with *ad-scandere*, a word that includes *elevation/ascension* in its meaning. Thus the original *askesis*, training, or physical exercise, begun at some point in Western history to denote spiritual exercise and practices,[33] aimed at attaining true perfection through detachment from the world and elevation to God.[34] Through this complex process, lightness has also acquired moral connotations. Lightness is traditionally associated with the pursuit of worthwhile goals and with moral goals of spirituality and elevation to God.[35] Not coincidentally, fasting has always been considered as one of the most effective *ascetic techniques*.[36] St Jerome prescribed mortification of one's body "by abstinence and fasting".[37] Fasting causes *weight loss*.[38] Thus, when ascetic practices (practices of exercise) spread as a means to moral integrity, moral connotations started being ascribed to *lightness*. Lightness has begun to be connected with *spirituality*, and because spirituality has always been linked to morality, *moral* connotations have been attached to lightness. This has all contributed to reinforce the idea that the body has to be transcended, and that morality is all on the side of spirituality and rationality.

Food restriction, moreover, indicates self-government, discipline and the submission of the 'corrupted' side of the human being. Fasting has been (and still is) associated with ideas of control over the chaotic passions of the body, and the person who is able to exert control over hunger, such a powerful physiological impulse, has often been presented as an example of moral integrity.[39]

Moreover, fasting has been associated (and is still associated) with the idea of *purity*. Fasting is 'detox'; this of course means that eating is always, more or less, a form of pollution. Fasting 'cleanses' the organism. Being empty of food is being clean.

We may look at eating disorders and over-exercise in light of this. The pursuit of thinness is central to the diagnosis of eating disorders (see Chapter 1). This means that eating disorders are an *intended* search of lightness. The most obvious and primary weight control method is restriction of food intake; associated with it, we find compensatory practices are also called 'cathartic' practices. Having an empty stomach, being free from food, is being 'clean', being skimmed from fat indicates the absence of corpulence and that the spirit has won over the most pressing impulses of the body.

It is the value placed on self-control and austerity, and the role of fasting in achieving these, which is the dominant background of the psychology of eating disorders sufferers. People with eating disorders, including those who exercise to excess, deliberately try to lose weight on the basis of moral reasons. I am not claiming that this process is fully conscious, or that anorexics want to become light because they think, at a conscious level, that this will make them morally good. What I am arguing is that the deliberate weight loss that is central to eating disorders, and the deliberate pursuit of physical excellence that is central to over-training, become understandable if one considers this particular moral background. The pursuit of lightness and physical excellence at all costs – which may seem unintelligible and irrational – actually makes sense in light of the moral values of self-control, austerity, discipline and spirituality that are deeply rooted in Western culture, both religious and secular, and which have been incorporated into ordinary morality.[40]

Control of the body is a reason for pride. The body becomes the arena of a moral fight, in which the sufferer expresses her strength of character. The sufferer identifies the locus of morality in the fight against the body.

Research on eating disorders has shown that moral values such as self-control, perfectionism, responsibility, intellectual achievement, hard work – values that traditionally are a part of Christian ethics – are invariably internalized by people with eating disorders: eating disorders cannot be fully understood without reference to these values.[41] Sociological studies also show that this ethic is still dominant in contemporary society.[42] The way fasting and *detox* are often advertised as means to well-being are a further indication of how this moral background has spread in contemporary society.

Physical and spiritual purity

Anyone who can make a simple search on the internet will find hundreds of health farms that offer fasting as a form of detoxification and purification. Invariably, these suggest that there is a connection between emptiness (freedom from food) and purity (spiritual catharsis, freedom from stress and disease). Control of the body is physical-spiritual catharsis, demonstration of willpower, virtue and moral character.

The following advertisements present fasting and detox practices. It is interesting to note that fasting is not only presented as a method of physical healing, but also of spiritual enhancement and of freedom.

Box 5.1 Examples of detoxification programmes advertised on the internet

From 'Shirley's Wellness Café':[43]

> **Therapeutic Fasting and Detoxification – Internal Cleansing – Optimum Health Through Fasting and Natural Hygiene**
> Detoxification is a normal body process of eliminating or neutralizing toxins through the colon, liver, kidneys, lungs, lymph and skin. Fasting is the world's most ancient and natural healing mechanism. Fasting triggers a truly wondrous cleansing process that reaches right down to each and every cell and tissue in the body.

Most extraordinary is the following case study.[44] Tim reports his experience of fasting and its benefits, after having participated in the 'Hawaii Detoxification and Meditation Programme'. Allegedly, Tim underwent 49 days of fasting (remember that Jesus is said to have undergone 40): 31 days on pure water and 18 on restricted coconut water and pure water. He reports, in his own spelling:

> [. . .] unparralleled experiences of inspiration, hope and beauty [. . .]. Each day was so amazingly different from the previous, so stock full of wonders, but each, in essence, consisted of lying down, drinking water, going to the toilet and sleeping.

> I won't relate exactly what happened to me. Nothing that I expected to happen, happened; the things I least expected to happen, did. If you hear my stories, you might expect the same things to happen to you. I can tell you of my tongue's coat changing and parting like the red sea, moles falling off, grey hairs vanishing, warts disappearing, knees and entire leg bones straightening, eyesight going wild then improving, a little toenail appearing for the first time in my life, [. . .] And who knows what miracles occurred in my kidneys, my spleen, my prostate, etc., etc.? [. . .]

Generally though, I am amazed at the overall rejuvenation that has occurred. My hair and eyes are shining. My brain feels like it is being fed properly (for the first time) – in essence, I am substantially cleverer. I am calmer. I am happier. I am wiser. My skin is as taught as rubber. I sat next to a super-handsome 19 year old Olympic triathelete in the plane home and the stewardess asked if we were brothers as "we looked so alike". (I just had to tell you that.)

Tim

Tim attributes all these physical and mental benefits to fasting. Tim even published some photos of his tongue, which, he declared, significantly improved after the 31 days of water fasting. We are told how Tim's tongue is clearing between weeks 5 and 6. (It is probably necessary here to provide a health warning: I strongly discourage anyone from following any such practice.)

These advertisements reveal a mixture of different elements, some of which may help to understand eating disorders. Through fasting the person achieves 'purity': physical and spiritual. It is possibly this purity that the eating-disordered exerciser is seeking to achieve.

Another worrying practice is so-called water therapy for the colon (colonic irrigation), a kind of enema, but more invasive, that is thought to cleanse the whole intestine. In one advertisement, we read:

"Whereas an enema (which you can do yourself) bathes only the lower portion of the colon, colonic irrigation (which must be done by a trained practitioner) attempts to clean the entire – roughly five-foot – length."[45]

Some so-called 'medical' societies advise carrying out enemas at home, while recommending that children and elderly people do not perform enemas more than three times per week. They advertise enemas with various substances, for example coffee, and recommend that in case of cancer, the dose of enemas should be increased to four times per week. There are hundreds of farms or apparent medical centers offering this potentially harmful practice as a way of healing and detoxing. A web search will show this. This seems to indicate the success of these farms.

The American Cancer Society has explained that: "Available scientific evidence does not support claims that colon therapy is effective in treating cancer or any other disease. Colon therapy can be dangerous and can cause infection or death."[46] Eating disorders and other abuses of the body can be understood in this context. People with eating disorders do not only want to be thin: they want to be empty and clean of food. Food is a pollutant, an undesirable intruder that makes you bloated, dirty and heavy.[47] One anorexic said: "Before I eat (or ate) I felt afraid that I had held out too long; while eating my main idea was how I could get rid of the food in one way or another – and this thought filled my head until I felt empty again."[48]

Through control of hunger and body shape, the person affirms his or her *purity*. Eating disorders and over-training are the consistent implementation of a morality that places high status on self-control and willpower. This is why the more emaciated, frail and vulnerable the sufferer becomes, the more powerful s/he feels.[49]

The claim that fashion may be responsible for eating disorders and/or phenomena of over-training is, thus, at best simplistic. The preference for super-thin or very muscular bodies stems from values that have an historical and moral substrate, and which are not merely aesthetic. Fashion is an expression of these values, similar to eating disorders and other forms of abuse of the body.

Why are eating disorders a Western phenomenon?

It could be asked why, in other cultures and religions placing high value on the spirit, such as Hinduism and Islam, eating disorders and fitness fanaticism are not so widespread. Similarly, it could be asked why – if all of us are influenced by the same morality – it is that these two phenomena are mainly found among females and higher sociocultural groups. Does it mean that males, for example, are less receptive to moral values?

Mervat Nasser has provided a comparative study of eating pathologies in different cultures, and I therefore refer interested readers to her work. She has shown that eating pathologies are indeed more frequent in high sociocultural groups, and are increasingly present in non-Western countries.[50] The fact that a higher prevalence of eating disorders and abuses of physical activities is found in Western countries and among specific sociocultural groups does not mean that the analysis offered is mistaken. Clearly, not all those who place high value on the spirit will develop anorexia. Moreover, eating disorders are probably multifactorial conditions (see Chapter 1). A discourse on the underlying values can thus offer a perspective from which these phenomena can be understood.

Conclusions

In Chapters 3 and 4, and in this chapter, I have looked at the possible causes and explanations of eating disorders and over-exercise. This chapter, in particular, has assessed a widespread argument, according to which eating disorders and over-exercise are results of social pressure. I have argued that moral values lie behind the aesthetic values, and that therefore blaming the media or the fashion industry is somehow falling short of the full explanation.

Tackling the moral values and beliefs that lie behind some abuses of the body does not mean advocating a policy of moral anarchy. It does not mean that, in order to resolve eating disorders, we should give up any moral belief and con-ception of what is good and bad, or right and wrong. It means, instead, that in the management of eating disorders, this moral background needs to be understood and addressed.

Although the media is not, strictly speaking, responsible for eating disorders and other abuses of the body, there is scope for legislation in the regulation of the portrayal of women and men in the media in general, and in the fashion industry in particular. This can have a positive impact on people's body satisfaction and correct one of the factors that can precipitate eating disorders.

But in order to resolve eating disorders, one has to look beyond aesthetic preferences, and understand why people identify self-worth with control, and, more particularly, with body control. It is this core value that needs to be addressed. The body must be transformed from an arena of moral combat, into a part of the self to be respected and loved.

The fitness profession can play an important role in this transition from fight against the body to self-love. Those who have attempted to reassure an anorexic by saying 'how thin' s/he is, will have experienced definite defeat. Saying to an anorexic that s/he is thin or beautiful will be of no reassurance to her. Beauty is not what the eating-disordered client is looking for. S/he is looking for value. S/he does not want to be beautiful: s/he wants to be worthy, as Princess Diana also suggested (see the beginning of this chapter; see also Chapter 1 for more on the psychological dynamics involved in eating disorders). She does not need aesthetic reassurance, but reassurance about him/herself as a person.

The fitness industry can also have an important role in correcting widespread and potentially harmful beliefs around the alleged beneficial effects of programs of detoxification, fasting, water purifications, and other forms of body abuses. These messages create confusion in people, with or without eating disorders. These practices might be sometimes used by eating-disordered sufferers because they might not know how harmful and ill-grounded they are. Thus, simply correcting false beliefs can be crucial to the fight against eating disorders. There are many ways in which the fitness profession can intervene in this arena. On the one hand, training courses for professionals should incorporate information about pseudo-medical information. The fitness bodies responsible for the education of fitness professionals should, in other words, ensure that instructors are competent to give proper information on the practices discussed in this chapter, and on similar potentially harmful practices advertised on the web or coming from dubious sources. Fitness professionals, on the other hand, could and should be directly involved in correcting false beliefs about pseudoscientific healing methods on sale off the web. Open communication with clients, informative posters in the recreational and exercise areas, advertisement of proper nutritional and health advice, organization of workshops or delivery of handouts for members, are all ways in which the fitness profession can reduce the negative impact of allegedly healthy advice on people.

Fitness instructors can, and should, encourage people to adopt healthy behaviors, without unnecessary and counterproductive moralization and stigmatization. Disapproval (blaming them or making them feel 'wrong') is unlikely to produce positive results. It is like trying to manage obesity by making the person feel 'wrong' about being fat: they already feel wrong enough in themselves. Fitness and sports professionals can, and should, encourage people to appreciate the benefits of being healthy. Having eating disorders is neither proof of morality nor proof of immorality

– it is simply bad for you. It's not bad *of* you, but *for* you, and this is the most simple and effective message that fitness professionals can get across.

Fitness professionals are a category of professionals who work on people's bodies, and that have, among their primary functions, the role of enhancing people's health. Therefore, they are not only challenged to tackle, but also stand on a privileged platform for tackling, the issue of eating disorders and over-exercise. A clinical psychologist who is director of an eating disorders unit in Baltimore, Maryland, once told me that it is fitness professionals who have the most power over the eating-disordered person. "If I say to my patient that she shouldn't miss breakfast, she won't listen to me. But if the instructor tells her to eat before a training session, she'll eat."

Key points

- The media alone is not the cause of eating disorders and over-exercise.
- Sociocultural changes partly explain the preference for thin and toned bodies.
- Ordinary moral values are to some extent responsible for eating disorders and other body abuses.
- It is often fruitless to try to reassure an eating-disordered exerciser by saying that she is beautiful or thin.
- Eating-disordered exercisers want to be worthy, not just beautiful.
- Fitness professionals should correct mistaken beliefs around the benefits of dubious cleansing practices (fasting, hydro-colonic therapy and others).
- Fitness professionals can have significant influence over eating-disordered exercisers, but it is important that they understand the basic psychological dynamics involved in eating disorders if they choose to supervise eating-disordered clients.

6 Exercise and eating disorders

Introduction

In November 1993 some photos of Princess Diana exercising in a gym were surreptitiously snapped and published in the *Sunday Mirror* and in the *Daily Mirror*. The Princess of Wales' lawyers started legal action over these photos.[1] Princess Diana was afflicted by eating disorders, and she was also a keen exerciser. Like her, many young, successful, bright people are afflicted by this condition and exercise, often to excess. Exercise is a weight-containment measure. Not only does exercise burn calories and fats, it also improves cardiovascular fitness and muscle mass, which affect basal metabolic rates (in other words, regular exercise increases our capacity to expend energy at rest[2]). However, eating disorders sufferers are, by definition, debilitated by unbalanced, irregular or scarce nutrition and by the use of cathartic practices such as self-induced vomiting and abuse of laxatives and/or diuretics. Whether exercise can be beneficial to such people, and in what way, is dubious.

So, imagine that Diana had wanted to enroll at your gym. But suppose you knew she was afflicted by eating disorders: what would you do? Would you pretend everything is fine, and thus expose her to preventable risks? Or would you raise the issue with her? And if so, how? Where would you find the courage, or the right words, to approach her over this?

This fictional example helps us to understand how uncomfortable it might be to make a decision about an afflicted person's training – whether they are a celebrity client or not. Certainly, their entrance into the fitness and sports environment raises important ethical and legal issues. Under what conditions should these vulnerable clients be allowed to exercise in a structured context (for example, a gym or group exercise)?[3] Is it ethical to allow people to take unusual risks during supervised physical activities? Who would be responsible if harm came to the client? For example, if a known bulimic had a cardiovascular arrest during an exercise session, would the sports and fitness trainers be responsible?

The legal responsibilities of fitness professionals will be examined in Chapter 8.

In order to decide whether or not people with eating disorders should be allowed to exercise, and thus, in order to give an unequivocal answer to the ethical dilemmas raised by the participation of eating disorders sufferers in sports and physical activities, it would be necessary to know the risks and benefits of exercise for them. Is exercise genuinely dangerous for eating disorders sufferers? Or can it also be beneficial? Should exercise be avoided altogether, or can some forms of exercise be beneficial to eating-disordered people?

A systematic study of the risks and benefits of exercise for eating-disordered exercisers has not yet been performed. This is probably due to the important ethical and methodological difficulties that any such study would encounter. First, sufferers are unlikely to disclose their problem. It is thus potentially difficult to collect people who are willing to participate in a long-term study of this kind. Second, and more importantly, there are ethical issues associated with such a study: sufferers would be knowingly exposed to preventable risks for the sake of scientific knowledge. An alternative way of studying the effects of exercise for eating-disordered people would be to modify exercise to make it safer for the subjects: the results of a modified study, however, would not necessarily inform us of the risks and benefits of normal sports and fitness activities.

In this chapter, I explore what types of exercise are more appealing to eating disorders sufferers and why. I give an account of our bodies' system for producing energy, in order to clarify why the eating-disordered exerciser is particularly at risk. I will discuss the benefits of exercise for the general population, and consider whether these benefits might also apply to the eating-disordered person. I will also provide an account of existing studies on the relation between sports, exercise and eating disorders.[4]

Energy systems

The body, whether at rest or while moving, requires energy. At rest, energy is needed for the body's chemical reactions to take place. When moving, energy is required for contraction of the muscles.[5] Energy is produced by the breakdown of high-energy phosphates. The main high-energy phosphate stored in the muscles is adenosine triphosphate (ATP). The stores of ATP are limited, and after a few seconds of effort they have to be resynthesized. This can happen in three ways, and there are thus three energy systems: one utilizes oxygen, and is thus named an aerobic energy system ('aerobic' literally means 'with oxygen'), the other two do not and are thus named anaerobic energy systems (anaerobic literally means 'without oxygen').

The anaerobic systems assist us in sudden bursts of energy (a sprint, for example, or lifting a heavy weight), or at the beginning of strenuous (or maximal) exercise. In the earliest or most strenuous phases of physical activity ATP is created through the synthesis of creatine phosphate (CP). This system, sometimes called the *phosphocreatine system* (PC), produces energy for a short time (around five seconds).[6] Here, glucose or glycogen present in the muscles is used to produce ATP through a series of complex enzymatic reactions. "Near the beginning of exercise, when oxygen consumption is low, and during intensive exercise when pyruvate[7] production exceeds the capacity of the aerobic energy system to oxidize it, excess pyruvate is reduced to lactate."[8] This second anaerobic system is called the *lactate system*.

The aerobic system assists us in activities of longer duration at lower intensity. The aerobic system can utilize glucose (aerobic glycolysis) or fat. Aerobic glycolysis assists the body in moderate physical activities, for up to two hours, unless carbohydrates are replenished.

The three energy systems normally interact. However, one specific activity can be *predominantly* anaerobic or predominantly aerobic, depending on the duration and intensity of exercise. For example, at the start of exercise, when muscles are not yet irrigated with oxygenated blood, the anaerobic systems will contribute to the vast majority of energy produced. After the body is warm, which could be five to ten minutes after the beginning of exercise, the aerobic system starts contributing to the energy production. After approximately 30 minutes of exercise at low/medium intensity, fat will be used to produce energy, whereas the other systems will assist to a lesser degree. However, if during a moderately paced walk or jog (which will be predominantly aerobic for many averagely healthy individuals) we encounter a steep track, then the extra energy required will be provided by the other systems, and usually by glucose.

The interaction of these systems will also depend, in addition to intensity and duration, on the level of fitness and genetic constitution of the exerciser. People with greater cardiovascular capacity (this is the capacity of the heart to deliver higher quantities of oxygen to the muscles for an extended period of time, and the ability of the muscles to use oxygen as a fuel)[9] have higher anaerobic thresholds; it means that their heart is more efficient in pumping oxygen to the muscles, and the muscles are more efficient in utilizing it to produce energy: they will thus be able to perform higher intensity activities for longer.

Benefits of exercise and fitness

Exercise has long been known to be highly beneficial. Pioneer studies in the 1940s began to highlight the benefits of physical activities for human health.[10] Well-being connected to physical activity is generally called 'fitness'. I will now look at some definitions of fitness and at the major benefits of exercise on physical and psychological health.

Fitness is typically associated with a state of wellness and with a variety of 'good things'. It is also associated with a given purpose: 'fit for X'[11] (I might be 'fit' but not 'fit for a marathon'). But 'fitness' is a rather elusive concept. One of the first authors to talk about fitness described it as "the ability to handle the body well and the capacity to work hard over a long period of time without diminished efficiency".[12]

The American College of Sports Medicine (ACSM) makes a distinction between *health-related physical fitness* and *physical fitness*.

Health-related physical fitness is defined as "an ability to perform daily activities with vigor and demonstration of traits and capacities that are associated with a low risk of premature development of hypokinetic diseases" (these are diseases related to or caused by reduction of movement, which is normally associated with ageing).[13]

Physical fitness is defined as "a set of attributes that relates to the ability to perform physical activity".[14] Physical fitness has, according to the ACSM, five major components (see Box 6.1).

Box 6.1 Components of fitness[15]

1 *Muscular strength* is the force or tension that can be exerted by a muscle or muscle group against a resistance at maximal effort;
2 *Muscular endurance* is the ability of the muscle to exert force or tension against a resistance for an extended period of time, at sub-maximal effort;
3 *Cardiovascular-respiratory capacity* is the capacity of the heart to deliver oxygen to the muscles, and of the muscles to utilize it efficiently;
4 *Flexibility* is the capacity of a muscle to extend to the whole range of movement around a joint;
5 *Body composition* is the ratio of fat to lean tissue in the body.

Subcomponents of physical fitness are listed in Box 6.2.[16]

Box 6.2 Subcomponents of physical fitness

1 *Speed*: the ability to move quickly and efficiently. Flexibility, co-ordination and core strength are essential;
2 *Power*: the combination of speed and strength;
3 *Agility*: the capacity to change direction at speed;
4 *Balance*: the ability to maintain equilibrium;
5 *Coordination*: the ability to move the limbs precisely in a desired direction.

The benefits of exercise are obviously dependent on the type of exercise chosen. Fitness is, in fact, *specific* to the chosen activity: for example, flexibility exercise will improve flexibility and not strength, and vice versa. *Specificity* is one of the principles of fitness.

Broadly speaking, exercise has long been known to benefit our cardiovascular system and body composition. Physical activity is one of the factors that are proven to delay death.[17] Physical activity decreases the risk of developing many health problems, such as atherosclerosis, back pain, some types of cancer, chronic lung disease, coronary heart disease, diabetes, hypertension, mental health problems, obesity, osteoporosis and stroke. Exercise thus reduces premature mortality, but also increases quality of life and independent living.

Box 6.3 offers a summary of the benefits of fitness.

Box 6.3 Benefits of fitness[18]

The benefits of fitness include:

- improvements in aerobic capacity (the 'fit' person can perform the same amount of work with lower heart/respiratory rates and lower systolic blood pressure);
- retardation of the natural aging process in aerobic capacity, loss of muscle and loss of bone mass;
- reduction of anxiety levels;
- increased level of body satisfaction;
- lower body weight;
- lower percentage of body fat;
- lower systolic and diastolic blood pressure;
- lower total cholesterol;
- consequent lower risk of cardiovascular diseases and Type 2 diabetes;
- increased bone mass;
- consequent prevention of osteoporosis (this is mainly produced by weight-bearing activities);
- maintenance of pain-free movement (this is mainly produced by flexibility, as being constrained to move within a subnormal range of motion is associated with pain);
- regulation of body weight (I discuss this later, in the account of obesity).

Muscular strength and endurance can be obtained with different types of exercises and physical activities, and benefits will vary accordingly. However, *muscular strength* generally contributes to increased resting metabolic rate, increased bone mass and consequent reduced risk of osteoporosis, increased glucose tolerance, joint integrity and improved posture.

In cases of anorexia, muscular strength is an obvious concern. As we have seen earlier in the book, muscles are used to provide energy, when fat stores are depleted. Lack of menstruation is associated with mineral bone density loss. Training that focuses on muscular strength might thus counteract some of the effects of extreme dieting. What I say here is not an invitation to take on an anorexic and put him or her on strength training. As I will explain in Chapter 8, when the exerciser appears too debilitated to take any exercise, s/he should be asked not to exercise. I will also suggest that fitness professionals should be specifically qualified to train people with eating disorders and work together with doctors. I am giving an indication of the components of fitness that might be jeopardized by abnormal eating, and suggesting how, in principle, exercise might be used to reverse some of the side-effects of eating disorders.

Cardiovascular-respiratory endurance has a number of very important benefits. These include reduced blood pressure, reduced body fat, reduced heart rates, higher lactate tolerance (this means that the body can be subjected to higher intensity work before lactic acid is produced as a waste product), and increase in good cholesterol (HDL). These all contribute to reduce the risk of atherosclerotic diseases such as coronary artery disease and stroke.[19]

The heart of the eating disorders sufferer might be weak (see next section). Bradycardia could be a sign of weakness in the heart, in people with eating disorders. Even people at normal weight who use compensatory practices can be at risk of cardiovascular disorders, due to possible electrolyte imbalances.

Flexibility counteracts some of the processes of ageing, especially on the joints and length of muscles. Flexibility helps maintain a full range of motion. A decreased range of motion is one of the causes of poor motility in the elderly. There is no particular association between disordered eating and flexibility. However, some authors suggest integrating flexibility exercise in the training of anorexic patients.[20]

Body composition should be considered not just a component of physical fitness, but an indicator of health. High percentages of body fat over lean tissue indicate obesity which is related not only to poor physical fitness but also to ill health.

It should be noted that *coordination and balance*, subcomponents of physical fitness, are also essential to fitness and health. Good balance and coordination counteract one of the consequences of ageing: the decreased capacity of proprioceptors[21] to provide information to the central nervous system as to where we are and how to move effectively through space. Deterioration of proprioceptors is one of the reasons why some elderly people become 'clumsy'. Lack of balance and coordination (thus increased risk of falls), combined with physiological reduction in mineral bone mass density can be significant causes of fractures in the elderly. Maintaining good levels of balance and coordination through specific physical activity can reduce the risk of falls. It would be interesting to study the correlation between proprioceptors' health and eating disorders.[22]

I will now consider in greater detail the risks that exercise poses for people with eating disorders. Whereas a certain degree of risk is associated with any activity we do – indeed lack of physical activity is overall much more risky than exercise – exercise can be particularly hazardous for people with eating disorders.

Exercise and risks

Broadly speaking, there are three types of risks relating to exercise:

1   Casual risks. Any physical activity, even walking to the shops or using the stairs, involves some risk. 'Accidents' are usually completely casual, and

although a good level of general fitness can prevent these from happening, it is possible that some of these will occur while exercising.

2   *Inherent risks.* Other risks are sports specific. These involve mainly musculo-skeletal injuries. The most frequent of these result from continuous micro-trauma or overuse.

3   *Relative risks.* These are risks that are not due to chance nor to the type of activity performed, but to the physical condition of the subject. Sudden death should be included in this group. Sudden death while exercising is an extremely rare event and 80 to 90 percent of sudden death happens to people suffering from some form of cardiac disease.[23] The risks run by eating disorders sufferers while exercising can also be included in this category.

Exercise, eating disorders and relative risks

The risks associated with exercising in combination with abnormal eating regimes may be serious.[24] It is important that those involved in the care and treatment of eating disorders, including psychiatrists, psychologists and fitness professionals, are aware that exercise can be dangerous for eating disorders sufferers, even if they have normal body weight.

Wilfley *et al.* report:

> Vigorous exercise can be a means of weight loss or one of several tactics used by the individual to counteract the ingestion of excess calories or deal with body image concerns. Intense fears of becoming fat may exist in people across all weight groups and body shape. [. . .] terror of being fat can cause some individuals to fall into the trap of excessively exercising while still falling short of the 'perfect body'.[25]

People with eating disorders generally tend to target aerobic exercise. Aerobic exercise is particularly relevant to them, because stored fats can only be burned through aerobic exercise,[26] and moderate to vigorous aerobic activities are the most efficient for weight-containment purposes. Eating disorders sufferers will thus privilege aerobic activities, most commonly running, swimming and cali-sthenics,[27] including exercise-to-music classes (ETM).[28]

Fitness activities – for example, group exercise and ETM – typically offered by sports and fitness centers are among the most appealing types of exercise for people with eating disorders.[29] These fitness activities are in fact increasingly popular and easily available,[30] so many younger sufferers, such as students, can afford to take a great deal of supervised exercise. Fitness activities often have a significant aerobic/cardiovascular component; they are usually affordable to the general public; they are often 'drop-in classes' (the sufferer can in principle tour around different health clubs and different classes, thus hiding eventual over-partici-pation); the relationship between the instructor and the participant is normally impersonal and the goals of the participant are often unknown to the instructor;

no medical certification is normally required; given the impersonal character of the class, the person can 'hide' in the group and hope to be unnoticed; in the group, the person with eating disorders, who is likely to be, to some degree, debilitated, may find extra motivation to sustain the intensity of exercise; and these forms of exercise are typically performed with music, which is an important motivator. For all these reasons, eating disorders sufferers often opt for fitness and ETM classes. These types of physical activities, however, can represent a threat for them.

The first important consequence of strict diet is a reduction of body weight.[31] When stored fats have been used up, and fuel cannot be provided by food, the organism will utilize muscle mass to produce energy. The result of this process is debilitation.[32]

A second consequence of low body weight in women is amenorrhea. As we have seen in Chapter 2, lack of estrogens linked to absence of menstrual periods causes bone weakening, even in the presence of hormone replacement.[33] This increases not only long-term risk of osteoporosis,[34] but also risk of fractures during exercise.[35]

Anorexic women with menstrual disorders have significantly lower cortical[36] (radius) bone mineral density. They might also have lower vertebral bone mineral density and vertebral trabecular[37] density. The most significant loss in spinal bone density occurs in the first three years of amenorrhea. A four-year follow up of women who had regained normal menses shows that when bone mineral loss is the result of amenorrhea, this cannot be reversed (unlike from loss resulting from oligomenorrhea – infrequent or irregular menses).[38] Bone mineral density loss is not only found in restrictive anorexics, but also in bulimic sufferers.[39]

Poor nourishment has negative effects on the heart.

> The diminished sympathetic tone consequent to starvation in AN [anorexia nervosa] is associated with bradycardia and lowered blood pressure. But what has also become evident is that more serious cardiac complications of starvation – such as arrhythmic disturbances, mitral valve prolapse, and congestive heart failures – are not rare. The perils of continued intensive exercise as [anorexia nervosa] progresses are evident in light of these potential cardiac problems.[40]

Heart disorders connected with abnormal eating may be fatal,[41] and the stress on the heart caused by aerobic and cardiovascular exercises can increase the risk of death for the eating disorders sufferer.[42] In both anorexic and bulimic sufferers, electrolyte imbalance, resulting mainly from vomiting, can produce abnormalities in cardiac function and cardiomyopathy. Thus the eating disorders sufferer can be at risk of cardiovascular collapse, even during exercise at seemingly moderate intensity, and consequently s/he can be at risk of brain damage and death.

Bulimia also causes tooth decay due to the acidity of gastric juices that are expelled with vomiting (see Chapter 2).

People with eating disorders show ventricular enlargement in the brain, presumably a consequence of malnutrition. Around 25 percent of them demonstrate anomalies on electroencephalographic tests.[43]

As we also saw in Chapter 2, hypothalamic abnormalities are also typically associated with eating disorders. Whether these anomalies are associated with higher relative risks during exercise is unknown. However, they remind us that even at normal weight, a person with eating disorders can be malnourished and have various health problems, in spite of apparently normal appearance.[44]

Eating disorders sufferers are more than averagely subjected to muscle-skeletal injuries. Some result from the loss of bone mineral density (see above). However, when muscles are not appropriately trained, joints are also at risk of stress.

Those who are extremely emaciated, or those who have adopted purging behaviors, in particular vomiting, over a long period of time, should be encouraged to avoid any vigorous activity. Their heart might be too weak to fulfill the demands of exercise. Their muscles' stores of ATP and other energy producing chemicals might be depleted; their fat stores might be depleted, and their carbohydrate consumption might be too low to supply sufficient energy to the working muscles. The bones might be too frail to sustain prolonged efforts, and thus it appears clear that the eating disorders sufferer is at high risk of injuries and even death. To say the least, exercise typically is not beneficial to this group in the same way as it is to the averagely healthy individual. With eating disorders sufferers, exercise should be monitored carefully, their blood pressure taken regularly, and those who vomit should provide electrocardiograms and blood for electrolyte examination.

The interviews reported in Chapter 7 show that many fitness instructors appear unaware of the perils associated with eating disorders, and of the extraordinary risks that these clients run. In addition to this, they face the great challenge of identification of those at risk.

Before I discuss issues of identification, I examine another group that might use exercise for weight-containment purposes: obese people.

Key points

- Many eating disorders sufferers exercise to excess, to reduce or keep their body weight low.
- Fitness has several benefits for health, both physical and psychological.
- Fitness reduces premature mortality.
- For the eating disorders sufferer, physical activity can be a hazard.
- Risks of exercising specific to the eating disorders population are fractures and heart disturbances.

Globesity: What is it?

Obesity is regarded by the World Health Organization (WHO) as an epidemic. WHO estimates that more than one billion adults across the world are obese. It has coined the neologism *globesity* to represent the global spread of this condition.[45]

Obesity is defined in various ways.[46] Ogden schematizes possible definitions of obesity:

Body Mass Index (BMI)

This is the most popular definition. Here obesity is defined by the relationship between weight and height. Although this is a relatively easy test to use, it does not consider the differences in weight: for example, muscle mass weighs more than fat mass, thus a body builder would be classified by this test as obese, even if his or her fat mass is appropriate. BMI also does not verify the location of fat, which is an important predictor of morbidity.

Population means

Someone is obese if his/her weight vastly exceeds the average. This definition is clearly imprecise, as it makes obesity context-dependent. Someone who could be obese, as measured in a population of underweight people, would not be obese in a population of overweight people.

Waist circumference

When waist circumference[47] is greater that 102 cm in men and 88 cm in women, weight loss is recommended. Regardless of the definition of obesity, excess fat around the waist (rather than in the lower body) is associated with greater risk of developing cardiovascular diseases and diabetes.

Percentage body fat

This is usually assessed by a skinfold thickness measurement using a caliper, an instrument used to pinch the skin at the back of the arm or on the back. A more technical but more precise way of assessing the percentage of body fat is bioelectrical impedance involving the use of electrodes. As water conducts electricity and fat is an insulator, the speed at which current runs will be an indicator of the mass of fat relative to the water.

In this book, when talking of the obese person, I will refer to the person whose BMI is 25 and above. In spite of its limits, BMI is the most frequently used indicator of obesity, and it is appropriate for the purposes of this book. I will not discuss the causes of obesity, but Ogden has offered a clear and comprehensive review of these. Particularly interesting is her account of the genetic causes of obesity.[48]

Mens sana in corpore sano

The Ancient Romans used to say that a healthy mind (*mens sana*) resides in a healthy body (*corpore sano*). Obesity is associated with a higher incidence of disorders, and psychological health is also dependent on physical health. The main conditions associated with obesity are hypertension, high blood cholesterol, Type 2 diabetes, coronary heart diseases, gallstones, kidney stones, cholescystitis and cholelithiasis, gout, osteoarthritis, obstructive sleep apnea and respiratory problems, asthma, some types of cancer (such as endometrial, breast, prostate and colon cancer), pregnancy complications (gestational diabetes, hypertension and pre-eclampsia), menstrual irregularities, infertility, irregular ovulation in women and psychological disorders (depression, eating disorders, distorted body image and low self-esteem).[49]

Obesity and exercise

Epidemiological data on changes in physical activity between the 1950s and the 1990s show a strong association between increases in obesity and decreased physical activity.[50] It also appears that on average the obese person exercises less than the non-obese. The Weight-control Information Network, one of the information services of the National Institute of Diabetes and Digestive and Kidney Diseases (NIDDK) reports that[51] "more than 50% of adults in the US fail to adhere to the recommended weekly physical activity for health[52] – this is a minimum of 5 times a week of 30 minutes of moderate intensity exercise".[53]

It seems, therefore, that one way of combating obesity is increasing physical activity. Exercise clearly burns calories; but also, as we saw above, it increases basal metabolic rates, and thus helps maintain weight loss.[54] According to some, exercise is also an appetite suppressant, but results of research point in different directions (others report that exercise increases eating, whether or not it decreases appetite).[55]

Although physical activity can prevent obesity, it is unclear whether and how it should be used to treat obesity, or to help the person affected by obesity to lose weight and maintain a healthier weight. According to some research, exercise is the best way to maintain weight loss.[56] However, exercise is not pain- and risk-free for the obese.

First, exercise might cause much discomfort for the obese person. They will tend to become fatigued sooner and at lower intensities, not because they are weaker, but because of the extra weight they have to carry around. This problem is not found in gravity-free sports, such as swimming. They may also have unknown damage to the knees and heels, which would make exercise very uncomfortable and even painful.

Moreover, many obese people are at risk of hypertension and coronary heart diseases. These need to be taken into careful consideration while exercising. Coronary heart diseases may cause heart failure. The weight that the obese have to carry is greater than what is natural for a person, and therefore the amount of effort required to complete a fitness workout is also greater than that required for a normal-weight person. The heart may be fatigued during even a moderate session. Exercise for the obese person, therefore, needs to be carefully monitored, whether or not the client is aware of any existing heart disease. Hypertension, which, as we have seen above, is one of the common effects of obesity, may also have adverse effects while exercising. In particular, there are forms of exercise that might increase blood pressure: for example, isometric exercise. Anyone with hypertension should avoid isometric exercise, and the blood pressure of an obese client needs to be taken before weight-lifting activities are prescribed.

Because of these problems, some suggest that obesity *should not be treated*. They point out that there is a strong association between *treatment for obesity* and psychological disorders, because inability to achieve targets can cause frustration, anxiety and depression. Second, ill health is generated more by shifts in weight rather than by obesity alone. Diet is associated with physiological changes, and therefore some claim that it is potentially dangerous. The paradoxes inherent in treatment for obesity have been outlined by Ogden:

- obesity treatment aims to reduce food intake and increase energy expenditure, but restrained eating and exercise can promote overeating;
- the obese person may suffer psychologically from social pressure to be thin [. . .] but failed attempts to diet or exercise like thinner people may leave them depressed and feeling they are a failure and out of control, which is also detrimental to health;
- weighing is central to most treatments of obesity but it is not always a benign intervention – it can contribute to low mood and self-esteem and changes in food intake;
- obesity is a physical health risk, but restrained eating and exercise may promote weight cycling.[57] Clearly, this is also related to adherence, which is a great challenge in promotion of fitness generally, and in fights against obesity in particular.

If people affected by obesity represent a significant challenge for the fitness profession, then people with eating disorders possibly represent a yet more difficult challenge. In fact, whereas the overweight or obese client is easily identifiable, it may be difficult, if not impossible, to identify the person with eating disorders, and in particular with bulimia nervosa. People with eating disorders may have normal weight. On the other hand, very thin people might not be anorexic. Moreover, whereas the person with obesity, if engaged in exercise, will do so primarily to improve their physical fitness and overall health, the eating-disordered participant might often exercise *in spite* of their ill health, and indeed, to the further detriment of their health. Finally, whereas it is possible to monitor the frequency, duration and intensity of exercise with many obese people, the eating-disordered sufferer will rarely cooperate with the professional in order to improve health. But how do we know that someone has an eating disorder?

How to identify the eating-disordered exerciser

The eating disorders sufferer can go unnoticed. In some cases, s/he will not be emaciated. In many cases, people will remain secretive about their eating habits and exercise routines. Moreover, eating disorders are not an *all or nothing* condition. There are borderline cases. People might have overwhelming concern with body weight and have inadequate diet, might exercise somewhat to excess, but might not satisfy the conditions for a full diagnosis of eating disorders. It might be very difficult to determine who is simply 'overly' concerned about body weight and shape, as opposed to those who, instead, have eating disorders and over-exercise. In spite of these difficulties, there are important signs that the instructor can look at (see also previous chapters).

First, eating disorders are sometimes characterized by rapid body weight shifts. Second, exercise routines might suffer sudden changes. It is reported that

> while many anorexic women were physically active people prior to the onset of their eating disorder, their exercising takes on an increasingly frenetic quality as the syndrome unfolds. [. . .] Like the dieting, the physical activity [. . .] becomes ritualised and rigid [. . .] the extreme denial characteristic of these patients may permit exercising to continue, sometimes even in the face of advanced emaciation.[58]

Over-participation can also be a sign of eating disorders.

Third, eating disorders sufferers might *understate*: that is, they might claim that they eat more than they evidently do, or that they exercise less than they evidently do. Fourth, in some cases, where the exerciser is burning out fat, sweat has a "distinct, foul odor".[59] Fifth, eating disorders sufferers might be affected by symptoms of over-training (as listed in Chapter 3).

The eating disorders sufferer who is over-exercising might also complain about being unable to sleep, feeling bloated or having an irritable bowel. They might suffer from anxiety, a frequent need for the bathroom (excessive urination, diarrhea), cramps, dizziness (resulting from low blood pressure) and muscle-skeletal overuse injuries.

Other signs of eating disorders could be[60] wearing baggy or layered clothing, appearing preoccupied with the eating behaviors of others, drinking soft drinks or water continuously, swollen salivary glands at the angle of the jaw, bloodshot eyes (especially after trips to the bathroom), preoccupation with calories and weight, repeated expressions of concern about being fat, high self-criticism about one's own body, consuming large amounts of food while remaining thin, bouts of severe calories restriction, eating in isolation or stealing food.

People at risk of having eating disorders are not always easily identifiable. But whereas this is true, and whereas there will always be people whose medical conditions will remain unknown to the instructor, this is no justification for not thinking about what one should do *when one has reason to believe that the activity they are teaching is likely to harm some of their participants.*

A consideration of the risks that the exerciser runs in the fitness premise is part of the role of the fitness instructor. Risk assessment and consequent action are legal and deontological (duty-based) responsibilities of fitness professionals (see Chapter 8).

I am not suggesting that fitness professionals should be overwhelmed by the concern of identifying every client at risk. *They should nonetheless be able to think intelligently and in an informed manner on how to deal with the client with eating disorders.* For this reason, they should appreciate the risks and potential benefits of exercise for eating-disordered clients.

Potential benefits of exercise for eating disorders sufferers

Whereas exercise might be contraindicated for people with eating disorders, according to some researchers sports and exercise might *prevent*, rather than trigger, the occurrence of eating disorders.[61]

A comparative study of male runners and female anorexics shows that the two groups share similar personality traits, especially the need for self-control, achievement and establishment of identity through control of the body. When the runners are unable to run they adopt disordered eating patterns similar to those of the eating disorders population. However, runners generally have higher self-esteem and lower rates of body dissatisfaction than their eating-disordered counterparts.[62]

This last finding is coherent with a well-established fact in sports and fitness: exercise and physical activities reduce body dissatisfaction, depression, anxiety, and therefore promote general well-being, not just physical fitness. Physical activity can provide a healthy solution for the psychological problems of people vulnerable to eating disorders or with low body self-esteem.[63]

In light of this, it is important to stress again that those who argue that exercise can cause people to develop mental problems, that exercise might be addictive and that it can cause eating disorders (Chapter 4), make a claim that has no solid evidence. In general, exercise is highly beneficial, and there is no sufficient reason to blame exercise or good nutrition if people abuse their body. The eating disorders sufferer is typically active even before the onset of the condition, and continues to be active after recovery.[64] Arguably, given the relationship between exercise and self-esteem, physical activity is, for the eating disorders sufferer, always affected by low self-esteem, and is more than just a way of burning off fat.

A supervised fitness training program[65] represents one example of how exercise can be used therapeutically and to help sufferers to achieve a better relationship with their body. Other researchers have also used physical activity in rehabilitation for eating disorders.[66] Of course, this implies that exercise has to be used within a monitored context, supervised by qualified personnel and tailored for the eating disorders sufferer.

A call for a new category of fitness professionals

Exercise has a good therapeutic potential for the eating disorders population. It can assist in improving self-esteem and body satisfaction, and it can counteract some of the secondary symptoms of eating disorders. Thus, in principle, it might be used in rehabilitation for eating disorders. Strength exercise and weight-bearing activities could be used in programs of prevention and correction of osteoporosis in those with long-term amenorrhea. Exercise could be used to help the sufferer to change her/his relationship with the body: from the arena of a moral fight, the body can become a part of the self to be loved and respected. Exercise can be turned from an instrument of self-destruction into an instrument of self-worth and positive self-regard.

This, however, requires the intervention of qualified fitness professionals who are well aware of the physiology and psychology of eating disorders, and are able to design exercise programs specifically for the eating-disordered client. I thus recommend fitness bodies responsible for the continued education of fitness professionals (like the ACSM, the UK Register of Exercise Professionals, or the equivalent in other countries) to design educational courses enabling advanced fitness professionals to supervise eating-disordered clients. The fitness industry proposes special modules that qualify trainers to teach wheelchair users, women pre- and post-partum, children or elderly people. On the same basis, professionals should be prepared to face the challenges posed by eating-disordered clients. Even at basic levels of education, fitness instructors should obtain information on eating disorders and on the essential psychological and physiological dynamics that characterize this condition.[67] This would enable them to recognize those at risk, and frame an action plan that could benefit the client.

One could object that eating disorders are a medical and psychiatric problem, and that fitness professionals should in no way be involved with them. Of course, I am not suggesting that fitness professionals should replace eating disorders specialists, but the reality is that fitness professionals are increasingly confronted with eating disorders. Since they cannot eliminate the problem, they should be equipped to handle it. In an ideal world, eating disorders sufferers would be supervised by specialized professionals, and perhaps would only exercise in purpose-built rehabilitation units. But in the real world, eating-disordered people go to the gym. In the real world, I qualify as a fitness instructor, and when I go to work I find a vomiter who might faint and return to the gym the next morning having eaten no breakfast. In the real world, I have an emaciated client who takes three-hour exercises one after another. And I do not know what I should do. If I talk to a colleague about this, I am likely to find that she or he has a completely different view as to what we should do (see Chapter 7). This is the real world and it is in this world that we need to work. It is this doubt and uncertainty that we need to address. The best way to address it is to enable fitness professionals to make informed and intelligent choices based on available evidence of the risks and benefits of physical activity for this group of clients, and make them aware of the legal and ethical obligations they have.

In addition to enhancing education, a scheme that allows easy liaison between fitness professionals and health care professionals could, and should be, encouraged. Within the UK's GP (general practitioner) referral scheme, doctors can refer patients to fitness centers or specifically to fitness professionals. The fitness industry could, and should, frame a similar scheme specifically designed for the management of eating disorders, liaising eating disorders specialists and fitness/sports centers. The possibility of liaison with an eating disorders specialist could provide vital assistance to fitness professionals, and also give the eating-disordered exercisers the option to decide spontaneously to seek the support of the health care professional.

Training eating-disordered clients without knowing what is involved in eating disorders and how exercise can benefit or threaten them might fall short of the duty of care to the participant (see Chapter 8). Given the growing presence of eating disorders sufferers in fitness and sports, it is imperative that fitness bodies intervene and give an adequate response to the call for professionalism and competence in this field.

Key points

- Recreational and supervised exercise can be beneficial to those with eating disorders.
- Exercise can improve self-esteem and body satisfaction.
- Fitness bodies are called upon to prepare fitness professionals to confront the growing presence of eating disorders sufferers in the fitness and sports arena.
- Fitness bodies should create routes of easy communication and interaction between fitness and health care professionals.

In the remaining part of this chapter, I will discuss one other issue related to exercise and abnormal nutrition: anorexia athletica.

Anorexia athletica

As already mentioned in Chapter 1, the term 'anorexia athletica' was first used in the 1980s by Smith[68] and Pugliese *et al.*[69] Originally, the term referred to the subclinical eating-disordered behaviors of athletes who, without reaching the full diagnosis of eating disorders, manifested some traits typical of the syndrome, such as excessive diet, purging and excessive exercise.[70] There is a higher prevalence of eating disorders among female athletes than non-athletes (up to 62 percent, compared to 3 to 5 percent in the normal population).[71] This is sometimes explained in terms of the pressure to excel, and of the emphasis on weight control as a part of the training regime.[72]

Anorexia athletica now has a broader meaning. It can be part of the behavior of the person who is excessively preoccupied with body weight and shape. Because anorexia athletica is not a disorder in itself (it is not included in the DSM-IV or the ICD-10), this term continues to be loose and to refer generally to over-exercise linked with an obsession with body weight and shape. See Box 6.4 for the definition used by the National Eating Disorder Information Center (Canada).

Box 6.4 Anorexia athletica[73]

Symptoms of anorexia athletica:

- Exercising beyond the requirements for good health;
- Being fanatical about weight and diet;
- Stealing time from work, school and relationships to exercise;
- Focusing on challenge and forgetting that physical activity can be fun;
- Defining self-worth in terms of performance;
- Rarely or never being satisfied with athletic achievements;
- Always pushing on to the next challenge;
- Justifying excessive behavior by defining self as an athlete or insisting that their behavior is healthy.

Who is at risk of anorexia athletica

It is believed that athletes, especially female gymnasts, cross-country runners, swimmers and dancers are more at risk of developing anorexia athletica. About 5 percent of young male athletes, typically wrestlers and cross-country runners, also display this pattern of behavior. Ballet dancers appear to be by far the most affected by eating disorders, and the literature on the link between ballet and anorexia nervosa is vast.[74] Up to 38 percent of female dancers in competitive settings are affected by serious eating problems.[75] In sports where thinness is thought to be essential to performance, eating disorders are present to a great extent: these sports are figure skating,[76] gymnastics and athletics. The ACSM reported that up to 62 percent of female figure skaters and gymnasts had eating disorders.[77]

Well-known in sports medicine is the 'Female Athlete Triad', a condition in female athletes characterized by disordered eating, amenorrhea and osteo-porosis.[78] Unhealthy eating patterns, purging and over-exercising can interfere with hormone secretion and lead to irregular menses (oligomenorrhea) or lack of menses (amenorrhea).

Amenorrhea appears in as many as 50 percent of competitive athletes, although a high prevalence of irregular menstruation is also reported among college women, and it is believed that menstruation is easily affected by emotional stress and not only by exercise.[79]

Is thinness really functional to exercise?

Whereas thinness can be seen to be functional to some types of sports and disciplines, excessive thinness is counterproductive and jeopardizes performance. As explained in Chapter 1, and earlier here, the person with eating disorders is *debilitated*. The anorexic in particular uses up muscle mass to lose weight, where fat stores have been depleted. This means that the excessively thin athlete will perform less well than athletes with adequate weight. Fatigue will result both from the loss of muscle mass and from iron deficiency (anemia), one of the consequences of poor nutrition. Electrolyte imbalances can be responsible for a lack of ATP which, as noted previously, is a source of energy.

Given that malnutrition weakens the heart (which like any other muscle, becomes smaller when food is insufficient), and given the enormous demands of competitive sports on the heart, the athlete with eating disorders is particularly at risk of cardiac arrest.

The absence of menstruation in the female athlete, also given the amount of pressure resulting from training for competition, puts the sportswoman at very high risk of fractures, which will obviously gravely inhibit performance.

The myth that thinness at all costs is necessary to performance needs to be discredited. A certain amount of fat, as well as healthy eating, are essential to best performance.

The role of coaching

Some believe that sports and exercise can precipitate the onset of eating disorders. This book holds the view that sports and exercise are not intrinsically harmful. They are intrinsically beneficial. Some people abuse them, and the challenge is how we should manage *these people*. Sports and fitness should not be demonized. Sports and fitness are one of the most striking expressions of human capacities. They can change people's lives for the better.

According to some studies, it is not sports themselves that lead the person to develop eating disorders, but coaching. It has been argued that coaches' comments about an athlete's weight can precipitate the onset of abnormal eating patterns.[80] Sports and dance, even at a professional level, do not lead automatically to body dissatisfaction and eating disorders. On the contrary, one study conducted by Ferraro found higher rates of *body dissatisfaction* among *the inactive people*. Ferraro stresses the responsibility of the coach in encouraging healthy body attitudes and eating behaviors.[81] I will emphasize this again in the action plan proposed in Chapter 10. Particularly heavy responsibilities fall on the coach of professional athletes and, by extension, all fitness instructors and exercise leaders. Also, early identification and intervention can prevent the precipitation of eating disorders.[82] Because early detection and intervention can help the sufferer to address the issue better, and prevent onset of a serious condition, it is crucial that eating disorders and over-training always be addressed and confronted as soon as they are identified.

Key points

- Anorexia athletica refers to over-exercise by those with eating disorders, and to athletes who manifest abnormal eating patterns.
- Eating disorders negatively affect performance in sports.
- Sports and exercise do not precipitate eating disorders.
- The coach has a crucial role in promoting appropriate exercise regimes.

Conclusions

Eating disorders are affecting an increasingly larger proportion of people in the general public, and sufferers are often drawn to fitness and sports centers. Fitness professionals must be prepared to handle the delicate situation of the sufferer. They can have an important role in the prevention and containment of eating disorders. Many of us encounter on a daily basis people with anorexia and bulimia, and we are all touched by the ethical dilemmas that they create. Fitness professionals, specifically, have ethical and legal responsibilities that need to be addressed and understood. The following chapters will present the problems, as fitness professionals see them, and provide legal and ethical answers to the questions that fitness professionals face. These answers, as it will become clearer in the following chapters, can by extension be applied to all fitness leaders and fitness enthusiasts. But, for now, let us turn to the stories that fitness professionals themselves have narrated.

Box 6.5 Tips for instructors: Identification kit

- The eating disorders sufferer prefers running, swimming and calisthenics – for example the gym's treadmill, step machine, stationary bike, cross-trainer and other cardiovascular training machines – and ETM.
- The sufferer might be physically active prior to the onset of the disorder.
- Physical activity becomes more ritualized and rigid as the eating disorder progresses.
- Exercise can be frenetic.
- Secrecy about eating and exercise habits.

- Rapid body weight shifts.
- Over-participation.
- Understatement – stating that they are not on a strict diet, when they evidently are, or that they exercise less than they do.
- Over-training.
- Wearing baggy or layered clothing.
- Bloodshot eyes after trips to the bathroom, which could be a sign of bulimic purging.
- Repeated expression of concern about being fat and high self-criticism.
- Consuming large amounts of food and remaining thin could be a sign of bulimia.
- Being overly concerned with the eating behaviors of others.
- Consuming high quantities of water/soft drinks.
- Swollen salivary glands at the angle of the jaw.
- Eating in isolation and on occasion stealing food.

7 People with eating disorders in the gym

Introduction

This chapter collects stories narrated by gym managers, gym instructors, personal trainers and studio instructors. Through their narratives, I will show how different categories of professionals are touched by the presence of eating-disordered exercisers, and how they deal with the challenges that such clients pose. The names of all people in the case histories have been changed so that clients and professionals are unidentifiable.

The experience of fitness professionals

Case history 1

M., gym instructor, personal trainer, football trainer

I work both as a gym instructor and as a personal trainer. I also work as a football trainer of a team of young boys, aged 15 to 18. I haven't had cases of eating disorders in the football environment, and in my experience I find that males are much less affected by these problems, because there is great social pressure on women to be thin. For males, pressure is to be hypertrophic, and you have to watch out for over-training and drug abuse, eventually. But even so, there are many more women with eating disorders than men who over-train or use drugs like creatine and anabolic steroid drugs.

I do one-to-one sessions, but I also design programs for clients as a part of my contract with the gym. I have cases of people with eating disorders literally all the time.

Me: How do you identify them?

In some cases, the least serious cases I'd say, they tell you what they eat, and you find that it is inappropriate for their age and shape. This is rare, however, because people with a deep problem generally don't like talking about what they eat. They feel judged if you find out. So, you have to pretend you don't know. Only those who have already one foot out of the problem might tell you the truth, if they trust you. But this is only once you have helped them to achieve a better relationship with their body and with exercise. In any case, it's clear when someone has got a problem . . . by the way they exercise, they don't listen to you and clearly overdo it. It's not normal to come to the gym in the morning *and* in the evening, or to do three hours in a row. It's clear there is something wrong with that.

Me: How do you deal with these clients? Have you ever refused a client because she or he has an eating disorder?

No, I am against that: people will get it elsewhere, you don't do any good to them if you tell them they shouldn't exercise. What I do is I design 'fake' programs. I mean, they are happy, they think they work out, but their workouts are far below the workout I would give to anyone healthy, not much in duration, because they like to stay in the gym for at least one hour, but in intensity. I insist that they should do aerobic workout, but I set up the machines for them in a way that their workout is minimal. If they say they are not fatigued, I tell them that that's because they are very fit. These people always need lots of positive feedback and gratification. I tell them we need to do a program that involves monitoring their metabolism. I assess metabolism, and if the metabolism slows down rather than improving, I discuss this with them and I make them aware and responsible for it. This is important for them psychologically, because if they don't train they feel guilty. This is what I call *non-training training*. If you tell them they mustn't come you have lost your client, you've lost it. You've got to get at them, and the only way to get at them is to give them the illusion of training.

Me: Is over-training ever an issue for you?

With eating-disordered clients, over-training is always an issue, because they want to over-train. So you have to do what I said before. You've got to make them believe they are training hard when they are not. Sometimes you have boys who don't have eating disorders but want to over-train. For example, I have a 16-year-old boy, Paul, he is one of my football team. In addition to what we do on the pitch, he now wants to join the gym. He has his training with me in the football team, and he doesn't need more training. So I told him I am not going to have him among my clients, if he comes here I am not going to follow him. In fact I know he has gone to another gym, but this is his problem. I told him if he gets an overuse injury I will kick him out of the team.

So, what's the difference between this guy and someone with eating disorders? Why is refusing someone with eating disorders not OK and not allowing Paul in the gym is?

You have to use your instinct somehow. A girl with eating disorders, if you tell her not to go to the gym, you damage her. You have to have lots of tact with them because they are psychologically fragile. This guy is not fragile, he is silly, and I won't comply with this. With him, I feel it is best to tell him he is being silly, push him on the pitch and make him work hard there. With the girls with eating disorders, I don't feel it's enough to tell them they are being silly.

Case history 2

R., gym instructor

I had one woman who was about 25 years old. She has been a client of this gym for years. At the beginning it was all normal, then she began to stay three hours in the morning and to attend two classes in the evening. She had normal weight, but she was somehow bloated. She wasn't fat, but she looked swollen. I can't really explain this, I could just see that there was something wrong with her. She fainted at the gym in front of me a number of times. Her blood pressure was very low. After some time, she told me she used to make herself sick, but I understood she meant this was still ongoing. We have worked together on this. As a gym instructor, this is easier than in the studio, because you can spend some time with the client alone, you can design a workout for her only, you can pace her, and take time to talk to her. Eventually, I managed to convince her that we should work on muscle mass and monitor the effects of exercise on mass and on her organism. I gave her nutritional advice, and explained that this was functional to the training we were doing. Over time, she got better. Now she trains less, much less, and her blood pressure seems more stable.

Case history 3

Gym instructor and personal trainer

One of the most touching stories I dealt with was one of the figure skaters. She was 16 years old. She was a figure skater, a good one, I mean, she got second at the European competitions. She was a very sensitive young lady. She was smart but very insecure. When she enrolled at the gym she wanted a personal trainer and we discussed her training routines on the skates, what she wanted to achieve, what she needed to improve, whether she wanted to work out for her performances or for other personal goals. By talking to her I understood that, despite her success, she was very frightened. As we worked together I got to understand her family issues: her sister was also an excellent figure skater, but she was smaller than her, and younger, and I think she felt her parents looked after her more, or loved her more. She was . . . kind of . . . 'already good' at everything, she was *taken for granted* . . . she was good at school, good at skating, she didn't need much support, whereas the younger got all the attention. I understood she had eating problems, and I kind of saw that these were somehow related to her feelings of worthlessness. I thought she was young, she was already having her training as a skater, but I accepted her because I thought I could and should help her, and I think it is my job to help people who have these sorts of issues. My program included training and nutrition. I made it clear to her that if she didn't follow my nutritional advice I would refuse to train her. She became a regular client, she followed my program, and we achieved good weight and good body satisfaction. When she first came she was thin, but not just thin, she looked dry and miserable, she looked unhappy. Once she told me that at school the teacher gave them homework, consisting of studying and talking about eating disorders, and she felt nervous about it. I advised her not to read anything about eating disorders, but to talk about her experiences, without telling the class she was talking about herself. I advised her to do this because I thought it would be good for her to express her feelings, to talk about her inner world. For me this client has been one of my greatest achievements. She resolved her problems completely over time. It was difficult because she was young, she was a girl and you have to be careful about young girls getting over-attached to you. You feel very emotional about them, but have to keep professional. And nobody teaches you how to do this, it has to come from within, and is hard. In the end, she stopped coming, she got back to figure skating only. I met her not long ago, she told me she wanted to go to college to study sports and exercise sciences. I tried to dissuade her, because she is brilliant, and I thought she'd be wasted in a career like mine – it's hard, you work odd hours, it's not very well paid. I thought she should study medicine. But she went on to the sports sciences college. She is in a relationship, she is back to normal. The one thing I understood is that the family plays an important role in these things. The family might control their kids, but often they don't touch the real issues. She was never gratified by her parents, and all her achievements in sports counted for nothing for her, because all she wanted was to feel important *at home*. When a daughter spirals down in that problem, one good mother or father will step in, will enter a relationship, but often the daughter's cries are neglected. Another thing I got out of this story is that we can really do very much for our clients. We have a power that maybe not even the doctors have. People with eating disorders sometimes admire us, they know we know about the body and about these things, and we can make a real difference if we play the right cards.

Let us analyze the choices made by these three professionals.

Case history 1 is a clear application of paternalism to fitness practice. M. uses *non-training training:* he tells 'a white lie'. M. is concerned with the outcome: he wants the client to get better. 'Paternalism' (from the Latin *pater* = father) literally means, according to the Oxford English Dictionary, "the principle and practice of paternal administration; government as by a father; the principle of acting in a way like that of a father towards his children". A person who behaves 'paternalistically' is, literally, a person who behaves like the father to his children. There is thus a positive meaning attached to this word. However, paternalism in ethics is more often associated with abuse of power, with the control over the life of another who is capable of owning his/her own destiny. The term 'paternalism' is wide-ranging, and many characterizations of the notion have been attempted both in philosophy and in medical ethics.[1] Moderate attempts to persuade the person, for example by talking to her and warning of risks, are not acts of paternalism. Excluding a person from the sports facilities, instead, *for her or his own good*, is an act of paternalism. Excluding her for the safety of others is not a form of paternalism. Telling 'a white lie' is an act of paternalism. Paternalism is thus used here to refer to interferences with someone's freedom in order to protect his or her welfare.

Case histories 2 and 3 illustrate attempts to establish a negotiation with clients. In both cases, the instructors set up the rules of the game, but make these explicit to the client: 'if you want me to train you, you will have to follow my advice'. In these cases, thus, the intervention of the instructor is not paternalistic. The client is informed, and can make a decision as to whether she wants to comply with the rules or not.

Case 3 is a particularly powerful narration. Here the instructor bases his judgment also on *how he feels*, on his instincts. "It has to come from within," he says. This capacity for intuition can be a precious resource for fitness professionals. The relationship between trainers and clients is primarily a human relationship, and as such has dimensions and layers to it that cannot always be rationalized and classified. Feelings are often shaky grounds for ethical action. How one feels about things does not often say much about how one *should behave*. People might have feelings that lead them to wrong others. Ethics is, in an important way, a rational exercise. Ethics explains why some courses of action are better than others. Looking at the pros and cons and strengths and weaknesses of various options can reveal how some feelings might be based on prejudice, or on ignorance or lack of reflection. Thus, certainly one should not purely rely on 'how one happens to feel'. Yet, in the delicate relationship with clients, what one feels to be right might indeed be right. Refining one's sensitivity towards the inner world of the eating disorders sufferer can be one of the keys to resolving the ethical dilemmas that arise in fitness. This is why understanding the psychology and physiology of eating disorders is important to determine the best course of action.

In case 3, the instructor establishes a meaningful communication with the eating-disordered client. He goes beyond what is said. He figures out what is happening to the girl. She suffers because the family is failing to give her the rewards she needs. Her eating disorders somehow relate to this. This enables him to provide the girl with some of the things she is missing at home: a sense of self-worth and of being valued by others. He also points out that fitness professionals can play an important role in the amelioration of eating disorders and over-exercise. I will return to this in later chapters. This is one of the reasons why it would be important to enhance, at all levels of sports and fitness education, knowledge and awareness of the psychological and physiological aspects of eating disorders, and to train professionals to deal competently with this widespread condition.

Case history 4

Val, gym manager

We had a case of eating disorders not long ago. It was a girl, and she was taking hours and hours of training every day. She was always cycling to the gym, cycling back, and you wondered how she handled all that training. She was very, very thin. At some point she lost her job. She was about 35, and losing your job at 35 for your eating problems is a big thing.

She constantly suffered stomach aches. She continually rushed to the bathroom. She said she suffered diarrhea. She used to stop coming for a month and then come back and resort to this routine of excessive training. At some point she was hospitalized for malnutrition.

Me: Did anyone ever approach her about this? Was she asked to leave, or to provide a medical certificate?

No, it was kind of impossible to speak to her. First of all, she was erratic: I mean, she would disappear for a month or two, then come back, and exercise like mad. If you tried to talk to her, she wouldn't talk to you, she was kind, but never really open to a true conversation. I tried, but felt it was like trying to get through a thick wall. She mainly went to spinning classes. And in the middle of the class she would rush to the toilet. It was miserable to watch, it really was. It was too big of a problem for me to deal with. Eventually she left.

Case history 5

Miriam, gym manager

We had the whole family at the gym: mother, father, her, and her brother. It was quite an unpleasant family to have around. The father was a dormouse, one of those men subjugated by the wife. The mother was one of those hypercritical and rigid mothers that are always unhappy about something their kids or husband do, or about the teacher, or about the machines, or about the showers or the weather. What I found very striking about this family was that the parents were really, really hyper-protective. Once that the girl's membership was expired, she told me that she should wait for the mother to come and pay, as she wasn't allowed to carry money with her. She was 21! I mean, to not allow your 21-year-old child to carry the money to pay off her membership at the gym is bizarre, I think. Her brother seemed OK, whereas over time the girl started training several hours a day, and losing a lot of weight. At that point I spoke to her parents. Then I asked one of our female members of staff to have a chat to the girl. I don't think the chat was at all beneficial to the girl. She kept coming regularly, she said she was trying to eat more, but she got thinner and thinner, and eventually she was hospitalized. When she came back, I had a chat to her, and she told me the whole story. I don't know why she decided to talk to me . . . of course I was informed, because the members of staff talk to me about what happens in the studio and in the gym . . . I mean, we don't see that this is a breach of privacy, we need to know what's going on with some clients. Anyway, maybe because I never stepped in before, she decided to talk to me about what had happened to her, her anorexia, her fear of becoming fat, her sense of impotence within the family (with which I could sympathize), and the hospitalization, she said, made her realize she was going too far. The parents were persuaded that they were controlling the girl far too much. They eventually sent the girl abroad for a few months, for travel, and I think this was an incredible achievement for them and for the girl. She was a lot better when she came back. She looked flourishing. Another thing that I think was crucial to her recovery, I think, was that the mother had a stroke. She understood she needed to grow up, take the lead, take responsibility. And she did it, she got things right. Now she has a job and a partner, she still comes to the gym, but her exercise is good now.

Case history 6

Hilary, gym manager

We have people with eating disorders all the time. Indeed, there are many people who come to the gym with different medical and psychological problems. However, I have had to turn down only one person in 15 years. The good thing about having a club is that we require membership, and don't do drop-in classes. It's a choice we make to have better control over our clientele. So, in a sense, we select our clientele in advance: we do proper health screening, if we think it is appropriate we request a medical certificate, and if the person is really unsuitable we can explain to them that they should not take exercise at this stage and we cannot accept them. But this has never happened, because we have qualified personnel who can work in rehab too, so we have had clients who had a history of cardiopathy, or diabetes, we have several members with asthma, and we work together with them. We also get GP referrals. I see our job as not just working for the fit and healthy, but also for the unfit and unhealthy, with some boundaries of course, for those who are acutely ill. Once people are enrolled, we work hard to keep them, for example, we organize events, like dinners out, on a regular basis. Clients get to know each other, become friends, and they love coming to the gym because it's a way of socializing. This helps us in our business, because we get to know people, and after some time they tell you about their problems, even if they haven't already told you at the health screening. In the case of that woman, she was just extremely emaciated, she was working out in isolation, riding a stationary bike for hours, or in any case using the cardio equipment for hours every day. I approached her, and said that my opinion was that she was over-training, and that her weight was dropping too low. She said this was her regular routine, and she was completely fine. Indeed, she was very regular and she never gave signs of dizziness or had any behavior that could induce me to think she was physically unwell. Yet, I was of the opinion that she was too thin to exercise at that level, and I requested a medical certificate. She brought it, and the certificate said she was "fit for general fitness activities". However, for me working out for three or four hours a day is not a general fitness activity. Athletes work out three or four hours, but general people don't and shouldn't if there is not a particular reason for it. Moreover, the certificate was *not* written by her GP, but by her gynecologist, who I happened to know was her sister-in-law! I regarded the certificate as invalid and I refused to renew her membership. I was very sorry, but my judgment was that any meaningful communication with this client was impossible, that she tried to deceive me with that medical certification, and I couldn't really let a person kill herself in my gym.

Cases 4 and 6 are examples of two completely different approaches to the problems posed by eating-disordered clients.

In case 4, the instructor opts for what in ethics is known as *respect for self-determination*. Attempts of communication (we don't know what types of attempts) fail, but the staff think that as she is an adult and seemingly capable of making her own choices, they should not interfere.

In case 6, instead, the environment set up by the managers helps professionals involved to get engaged with people's problems. The center has qualified personnel who can assist people with a number of disabilities and liaise with GPs. This helps to handle difficult cases. In this case, the manager believes it is not part of her job to allow a person to harm herself using the facilities. It is important to note that this is not a case of paternalism, as represented in case history 1. The professional is not acting against the client's wishes *for her own good*. The professional is acting against the client's wishes because she believes that supervising that client is not part of the fitness professional's job – it is not a part of what they are providing. This, as I argue in Chapter 9, is a form of *contractualism*.

I now give a brief account of contractualism, in order to explain the term and the way I use it in this context. Contractualism is a term of political philosophy. Among the historical fathers of contractualism, we should include Jean-Jacques Rousseau and Immanuel Kant, in the 1600s and 1700s. Contractualism is mainly a political theory, that is, it is meant to explain how civil society is generated. The core idea is that civil society comes into existence when individuals, from isolated beings, enter a *social contract* with one another. By doing so, they give up their state as individual beings, and become members of a whole. A society thus exists by virtue of that social contract that individuals stipulate with one another. The parties, in principle, are bound to follow the rules of the social contract. This contract is more often considered a logical construct rather than a real event, which helps to explain the existence of and to legitimize civil societies, with the constraints they impose on each individual for the benefit of the whole and of every individual as a part of the community.

Contractualism, in addition to being a political theory, also provides a way of thinking about moral actions. For example, for Rousseau an action is moral insofar as, within the social sphere, it is acceptable to others *as equal*. Thomas Hobbes had a different idea, which is called *contractarianism*. Hobbes believed that individuals act socially in their mutual self-interests. Tim Scanlon re-articulated this idea and argued that the basis for moral action resides in the rules that each rational agent has reasons not to reject.

Contractualism, apart from these important differences, considers morality as a result of the agreement made by those in the moral domain. It is in this broad sense that I use the term in this context. The argument that the moral entitlements and obligations towards clients should be derived from a proper understanding of the implicit moral contract among those in the moral domain is in this sense an argument of contractualism.

Contractualism and paternalism offer different grounds for action. To say that a client should not be accepted because accepting them means doing something that is not included in the implicit or explicit contract with the client is not a form of paternalism. Whereas the grounding motivation in paternalism is to do the person good, in contractualism the grounding motivation is that the terms of the contract are not respected, or that what is expected goes beyond the terms of the contract.

How we should understand the contract between fitness professional and client is not immediately clear. For example, in case history 6, the trainer believes that accepting the disordered client is not part of her job, and in case 3 the trainer believes that it is part of his job. Who is right? I will argue in Chapter 9 that the manager in case history 6 is right. Of course, working in fitness and sports also involves caring for the unfit and unhealthy, but there are important boundaries to set. Fitness professionals can care properly for the unfit and unhealthy only if they have the competency to do so. In case history 6 the manager sets up an environment that allows people with health problems to have proper training.

Since she has no personnel who can properly care for the anorexic client, she opts not to accept her. If someone is not qualified to deal with this difficult condition, and if exercise is a threat to health or life, it cannot be claimed that accepting the client is 'part of the job'. Of course it is not and, as I will discuss in Chapter 8, it can be a breach of one's duty of care.

Case history 7

Y., gym manager and acrobatics instructor

Y. owns her own gym. She is responsible for the acrobatics sessions. She teaches mainly girls aged 8 to 17 who are professional acrobats and compete at a national level. She also works for local primary schools. Here is her testimonial.

There are girls at acrobatics that suffer eating disorders. One is fully anorexic, but there are a couple more in the group who are overly concerned with their diet and tend to over-exercise. You can tell they dread fat and constantly talk about diet, workouts, how fat they are, how chubby their stomachs are, how chubby their thighs are.

Me: Are they really chubby?

You kidding? These are professional acrobats, they train three hours a day and they are young adolescents! They are slim, toned and they need to eat. They need to have some fat. If you put them on a diet you kill them.

Me: With these girls, do you speak to them? Do you ask them to stop training if they go too far? Do you speak to their parents?

We work very, very carefully with them. They are professional athletes, and they have to train about every day, so it's a difficult one. I generally tend not to speak to their parents. In over ten years, I never did. I think it's best to speak to the girls and build a relationship of trust and respect with them. If you involve their parents, they might feel betrayed. Especially at that age when they start building their own world and they begin to detach themselves from their parents. I take these sessions with two colleagues. We are aware of those who have the problem, because we spend time with them every day, and talk about it among ourselves, but we don't feel stopping them from coming is the right thing to do, even if we are aware of the risk they run, because they are still young, and we believe working with them is more beneficial to them than excluding them and 'punishing' them because they have an eating disorder. The one who is clearly anorexic, for example, has a terrible family situation, we would never abandon her. She lost her mother when she was only three, her father is with another woman and she doesn't feel loved and accepted by this woman. She is very attached to us, and if we reject her, she will hurt a lot. We have had her for years now, since she was maybe five or six, and I am also attached to her. It never really crossed my mind to tell her father to remove her, I just wouldn't do it. We work with her. Our main goal now is not to make a great professional athlete of her, although, if this is what she wants to become, we'll help her. Our main goal now is to help her to love herself. I think gymnastics can be very useful for that. I think we will manage to get her over it, together. Gymnastics can help you to love yourself. It's the coach who has to be good enough for that.

Interestingly, you know, I have the opposite problem with younger kids. At primary school we have an increasingly higher number of obese children, who don't want to take any exercise and just want to eat junk food all the time. We come here, and we have to fight to get older kids to eat, with younger kids we have to fight to get them not to eat and to exercise: that's another real challenge. They would spend their day at the PlayStation eating junk food. You have to really struggle to make it fun for them, because they don't like moving.

It is worth noticing the difference between this case and case 5. In case history 5, the professionals involve the parents. When the staff tries to speak to the mother, communication is ineffective. The staff might have failed for a number of reasons, but ensuring that the client, albeit young, feels respected is likely to promote a relationship of mutual trust and respect, which can benefit the vulnerable client. Minors, if mature enough, are entitled to the same respect.

The approach taken in case history 7 is an approach of care, and falls under the *principle of beneficence*. The principle of beneficence suggests that people have a duty to do what is good (*bene* in Latin) for others (this will be explained in greater detail in Chapter 9). The professional, in this instance, is concerned with the long-term welfare of the young lady. She is not purely acting 'as a trainer', but as someone who cares for the child.

In Chapter 9, I will argue that professionals are not morally obliged to care for clients in this way: it might be good if they do so, but there is no moral duty to do so. In ethics, actions that are morally laudable but not mandatory are called 'supererogatory'. As I will show in Chapter 8, in cases where the health and safety of the client is at imminent risk, the primary duty of care of the instructor is to stop them from exercising. In these extreme cases, there is no room for supererogatory actions. Duty of care requires suspension of activities.

Case history 8

Alex, gym manager

We have always had problems of eating disorders here. In any case in which I feel insecure about the person's health, I ask for a medical certificate. We of course do health screening, and also life-style screening, but if I am not 100 percent sure about the person's health, I request a medical certificate. When I know that one of the clients has an eating disorder, or if I have a reasonable worry about it, I don't necessarily refuse her access to the center. In fact, in general I don't refuse access to anyone. However, I speak to the instructors. If the client wants to do studio sessions, I will talk about my worries to studio instructors, otherwise, if the client wants to use the gym, I will inform the gym staff.

Me: Do you ever approach the person directly?

You can't really intervene in a strong way, unless you are the teacher, because it is a delicate situation, and there are many risks involved. There is privacy as well, and you can't really walk up to someone and say: I think you have an eating disorder, shall we talk about it?

Me: Would you refuse an application because you suspect the applicant has an eating disorder?

I could, but I would protect myself, not the client. I think our job involves helping the client. If you refuse an application, the client will enroll at another gym; there are plenty of gyms out there, and they will take her. I'd rather take responsibility than not.

Me: So you've never asked anyone to leave?

Only once, but not for eating disorders, it was drug abuse . . . hold on . . . I actually did. It was a woman, she had severe eating disorders, and also psychological problems. She was here for seven months, and she caused so much trouble. I don't know what she was taking but it was constant trips to the toilet. She would complain of dysentery. She was extremely skinny, she would attend spinning classes in a row . . . however many classes there were. If we had three classes scheduled, she would take them all. In the middle of the class, she would rush to the loo, complain about gastric pain, and then get back to the class. We had a meeting with the staff, and we discussed the issue. The clients had written a letter to me, complaining about this woman who would interrupt the class several times, and about how her presence would disturb the class. At the meeting, it was decided that this client should be asked to leave, and I had to write to her to explain the decision. She rung me, and she cried over the phone, she said that the gym was the only thing she had. I felt very guilty, but it wasn't a problem that we could resolve. It wasn't just a nutritional problem or an eating disorder.

Me: What about if you accept someone at the gym, and the instructor refuses to have her in her classes?

I have to respect the instructor's judgment. I would speak to the person and explain.

Me: How do you see a way out of this problem? We have a growing presence of eating-disordered exercisers; someone suggests that it spreads like an epidemic.

Yes, you are right. Twenty years ago, when I started my business, it was rare to see someone in the grip of an eating disorder. And it was only girls, only young, only from the middle and upper class. Now it affects men, older people, and those from a lower sociocultural background. They all come to the gym because it is here that they release their anxieties. It's a huge risk for us, a big problem. The way forward, for me, is education for instructors, even at basic levels. I am sure they do training on eating disorders at college or universities, but the instructors who take qualifications out there, they should be informed and educated, even at basic levels.

Case history 9

Alexandra, studio instructor

We had a girl, we still have her, and she suffered bulimia. I got to know this because she is a regular customer, and eventually she spoke to me. The key to the problem is trust. When people trust you, they speak to you, and you can help them. She told me that I was the only one to know. She gave me a big responsibility. She got over it, as far as I can see. She has a regular diet, and she seems fine.

The most touching story I had was of a girl, I mean, she was a woman, but she still lived with her parents. She must have been in her early 20s. Her parents were separated, but lived in the same house. She found this very disturbing.

She hated her father. She went on a diet, and she was called to model at a local show. It wasn't a big event, just a local festival, but it was after this that she stopped eating and started training three hours in a row. She was coming to my classes, and I really didn't know how to handle this. I decided to try and make friends with her, to get her to tell me about herself. On one occasion I invited her to a fitness convention, and she accepted. She was very enthusiastic about it. After that, I started slowly to explain to her that she would lose her muscles if she didn't feed herself properly, and she wouldn't be able to excel either in fitness or in any other area of her life. It was very touching. I was watching a beautiful, sensitive intelligent young woman – OK, there were family problems, but she had so many things to be happy about, I was watching her throwing her life away for nothing, watching her die, turn her life off slowly, day by day getting into a darker and darker tunnel. I was scared and felt impotent. I think she eventually admitted the problem and sought help, but it was a long, long way to recovery.

Case history 10

Ruth, studio instructor

. . . She was so thin, she was minute, short, tiny, she m1ust have weighed 25 kilos, and she was working out for hours and hours on the step machine.

She only used the gym, only on the step machine, for hours really. It was crazy . . . I mean, it would drive anyone crazy, can you imagine? Going up and down on a stationary step machine for three or four hours a day? One time I approached her, and while having a small chat I asked her: 'How do you manage to keep going for all these hours on just one machine? Don't you get bored?' She said: 'I just ignore it'. I shivered. The girl was asked to leave at some point . . . she was told she wasn't fit enough to do that amount of exercise, that they couldn't accept that responsibility, and she should seek medical help.

Me: Do you think the managers did the right thing?

Yes of course they did . . . I mean, we are not psychiatrists, these people need not to exercise, they could harm themselves badly . . . they . . . they are just too ill to exercise and it's not our job to care for them . . . it sounds harsh but that's the truth . . . we aren't qualified to deal with these issues. I feel that people like this want to force us to do something we are not fit or qualified to do, that is to take care of someone who is ill. She could end up with fractures, or even die, and we are not there to help people injure themselves. We promote well-being, wellness, we can't accept cooperating with someone who is killing themselves.

Case history 11

Claire, studio instructor

We had a client who was clearly anorexic. It was clear to everyone. She was somehow revolting to see. She had a very young body, but in her face she looked 50. Actually, she was only 26!

Me: How did you know she was anorexic?

It was obvious. She was a skeleton, and on top of it she was sort of . . . I don't know how to say, mad, frantic . . . one example . . . she would do step classes with three blocks . . .

Me: Can you explain what the blocks are?

The step is a platform, where you walk up and down on time with the music. To increase intensity of workout you can put two blocks underneath, one each side, which raise the platform by a few inches. If you are very fit, you can put four blocks, two each side, and this is the most that everyone would do. She was using three blocks each side. The platform was so high that she was jumping up and down, flying like a bird. I have never seen anything like this. When somebody that thin goes to the gym they must have anorexia, otherwise if you are that thin you are sick and you lie in bed. She was trying to sweat I think, but she couldn't because she had nothing to sweat out. She always had two bottles of diet coke on her. One friend told me this was her only intake for the whole day.

Me: Did she ever come to your classes?

No thank God she never did . . . I don't know what I'd have done. Probably I'd have let her do whatever she wanted. They banned her from the center, and I used to see her running around the city . . .

Me: Do you think they were right in banning her?

I can understand why they did it, but I think that was the wrong decision. This is no way to help these people. I think overall it was wrong to ask her to leave.

Me: And wouldn't you be worried that if she collapsed in your class you could be held responsible?

No . . . not really. To start with, I am insured. But also, if she collapsed this wouldn't be my fault . . . I mean, it wouldn't be because I did something wrong in the class, it'd be clear that it was because of how she was, and I think you can't leave someone in that state on her own. I believe we choose this job because we want to do something good for others, and kicking out people who are so poorly is doing no good.

Me: What happened to her, do you know?

Yes, she still lives in the area. She got worse. I think she will die soon . . . just by looking at her. She has a friend who every now and then takes her out for a walk or for a coffee. She has deteriorated since she was barred from the gym. She looks as if she is about to die.

Me: Do you feel it's the manager's fault, somehow, that she ended up like this? Or that the gym could have done something for her?

Yes . . . I mean, I am not sure, I wouldn't go so far as to say it's the gym manager's fault if she is like this. But it's hard to imagine that we could have damaged her any further. I think that if we maybe worked all together, we could have done something for her. I think leaving her on her own was wrong. We thought about ourselves, and not about her. She should have been kept in and looked after.

Cases 10 and 11 present a stark difference in attitudes. Ruth thinks that what the client is asking of them is not part of the proper understanding of the professional's role. The professional does not have the competence to look after people with eating disorders. Professionals cannot cooperate with the clients' wishes to harm themselves. She thus interprets the implicit contract between client and professional as one that has some clear boundaries, which are determined by the professional's competencies. The professional should remain firm to what he or she is qualified to do and attempt to do no more, especially when the health of the client is at stake. This is clearly a distinctly different attitude to other professionals, who seem willing to try and help in other ways.

Claire thinks in different terms. She believes that any harm that the client would suffer in the sports center would not be imputable to the professionals. The long-term negative consequences of barring her could outweigh the short-term benefits she obtains by being excluded from the gym. Claire is thinking in terms of minimization of long-term harm. We should prefer the course of action that minimizes the most severe harm. Chapter 9 will discuss the strengths and weaknesses of this way of thinking.

For the time being, it is worth noticing the distinct difference in attitudes. For Claire, "we choose this job because we want to do something good for others, and kicking out people who are so poorly is doing no good"; for Ruth, "we promote well-being, wellness, we can't accept cooperating with someone who is killing themselves". These are just two extremes, which highlight how differently people feel and think about their role as fitness professionals, and their duties and responsibilities towards exercisers.

Case history 12

Marie, studio instructor

We have seen many cases of eating disorders. I used to work as a freelance in quite a big fitness club. It was the only one in the area. Maybe this is because in this area there are quite a few unemployed people, so people feel worthless and they take it on to their body, she would always come by herself, I have never seen her around with a friend or a boyfriend . . . I don't know, but I have seen quite a few of these cases. Once the woman was barred altogether from the gym. She was good, she wasn't thin or anything, she was just very fit and she seemed to love coming to the gym. She was enthusiastic and good to teach. Every time I was teaching she was there. She was sculpted, she wasn't very young. Maybe she was in her late thirties or early forties. At some point, all of a sudden, she started losing lots of weight. You see when losing weight is a problem, because all of a sudden the person becomes thin, they lose weight very quickly, and they looked fine before. Their thinness looks unhealthy, they look dry, rather than thin . . . I don't know. She was still perfectly able to exercise, she seemed completely fine, she was smiling as usual. Her pants were just more and more loose. I spoke to my line-manager about it. My line-manager was also a fitness instructor and was the studio coordinator. She said she spoke to her about it. She admitted having a bit of a problem with herself. They didn't speak openly about eating disorders. My line-manager said to her she was 'going a bit too far' with her diet. She had to resume eating, otherwise she would be forced to ban her from the gym.

I didn't think anything at the time . . . I mean, whether she was right or wrong . . . I assumed she was right . . . I didn't know much about eating disorders, but I think if one gets too thin too quickly there's something going wrong. Well . . . in the end, she seemed to recover a bit of weight, but she relapsed, and she was asked to suspend activities. I don't think she was kicked out in a bad way. I think the line-manager spoke to her and tried to make her understand that this was for her own good, and that she would be welcomed back when she put on some weight. She came back in fact, after a while. Then the whole thing started again. She was barred again. I soon after moved away from the area, and I don't know what happened to her. But I will never forget about that one, because I have seen her going down. It was a pity, and we couldn't do anything for her.

Case history 12 will be discussed at length in Chapter 9 because, as I will argue, it represents an example of good practice.

Case history 13

Rita, studio instructor

I used to work as a fitness instructor before I had my child. I once had an anorexic woman. She was doing a massive amount of exercise and she was extremely emaciated. I spoke to the gym manager about it. He said I shouldn't worry, and it was good that someone so thin was coming to my classes, it would be good publicity for me!!!

Case history 14

Rosemary, studio instructor

I never happened to have very thin women among my clients, but I generally work with mature women, because I teach Pilates, and eating disorders don't seem to affect them as much as younger people. Or maybe people with anorexia would rather burn out calories and wouldn't necessarily go to a Pilates session. I only once had someone with a history of bulimia, but when she was coming to my classes she seemed fine. She wasn't thin, she was just well toned. After my manager told me she had had bulimia I thought it would be good if I gave her some positive feedback about herself. These people always have low self-esteem and need to be reassured. I always used to tell her what a nice figure she had.

Me: Weren't you worried something could happen to her?

No, she seemed fine. In any case, this is our job. Maybe you have a client with a history of infarction and they don't tell you. You do your screening, try to get as much information as possible out of them, and try to design safe workouts, but you can't force people to disclose their medical problems. All you can do is watch out that they are fine.

Me: So, you see eating disorders on a continuum with other health problems, and not as a condition that stands on its own?

No, they don't stand on their own. They are like many other medical conditions, people have them and don't tell you, and you can't force them. I see the problem is our job, not their disorder. We need to be extremely careful with everyone, even if they look fine and say they are fine.

Case history 15

Paula, studio and gym instructor

There are lots of people with eating disorders. Lory is one of them. When outside was 30 degrees she was taking liters of diuretics and missing her lunch and coming to the gym . . . then she was dizzy . . . .

> **Me: How did you know she wouldn't eat?**
>
> We all knew. But what do you do? How can you tell her *not to come*? I wouldn't do it, it's her choice. I know what she does is not normal, but many things are not normal, really. People who do three hours in a row . . . that is not normal either, but do you start deciding on what other people should do? It must be up to them. I mean, who am I to decide on who is coming and not coming to my classes? I can't stand there and say 'yes', 'no', 'you come', 'you go'. They have got their own minds and they have to decide for themselves. Lory knows she shouldn't come to my lunchtime class having eaten nothing since the night before, of course she knows that. I have told her that. She knows she has hypotension, and she knows that if she throws in diuretics she'll faint, but she's got her own mind, if she wants to take these risks I can't stop her . . . who am I to do that? In the worst cases, but I have never done this, you might request medical certification, then it is up to the doctor, it's a matter between the doctor and them, if the doctor thinks they can come to my class that's fine by me.

Further observations on gender roles

Before concluding this chapter, I wish to make additional observations relating to gender roles. The case histories reported highlight some important facts on gender roles, which can enhance our understanding of eating disorders and improve our capacity to deal with them.

All fitness instructors who have given their testimonials have talked about *women* sufferers. I have purposely avoided the question: "How many *male* clients have you had, or do you have, with an eating disorder?", because I wanted to see what would come to their minds, what experiences they found as most significant. Their responses highlight that many fitness professionals automatically think about eating disorders as a women's issue. This is probably because there are, in fact, many more female than male sufferers. It also appears that many fitness professionals are still unaware of the perils of eating disorders, and whereas humanely they feel inclined to help these clients, technically they do not raise the issue of what type of program should be tailored for them. Finally, it also appears from case 13 that some so-called 'sports and fitness' centers might even praise body abuses.

In case history 1, M., who also works with men both in the gym and as a football trainer, reports that men are rarely affected by eating disorders. Even over-training is, in his experience, more often a women's issue. M. attributes the higher prevalence of eating disorders among women to the social pressure to be thin. This pressure affects mainly women, whereas men are instead subjected to the pressure to be 'big' or 'muscular'. Where body dissatisfaction is found in men, this translates itself in a desire to 'bulk up' (hypertrophy). But men do not seem to suffer body dissatisfaction as often as women.[2]

This is of course only the experience of one personal trainer, albeit with years of experience in different fields of physical education. However, his observations are consistent with recent research on gender-related aspects of eating disorders.

In Chapter 1 I discussed the society of the eating-disordered sufferer. Eating disorders appear only in some societies, and are thought by many to be a culturally bound syndrome. Many have tried to identify traits that are specific to the societies in which eating disorders appear, and which could somehow explain or trigger eating disorders. Despite the increasing prevalence of male sufferers, eating disorders still mainly affect women.

According to Rosa Behar, "gender roles and sociocultural expectations appear strongly implicated in the development of eating disorders".[3] According to Behar, participation in sports and fitness activities that promote weight loss and modifications of body shape reinforce the tendency to develop eating disorders in women. The spread of eating disorders is, according to Behar, related to the modification of the role of the woman in society. This has been discussed previously (see Chapter 1). However, in addition to this point, Behar has also pointed out that in modern Western societies it is *still* in some ways required of men that they "be domineering, aggressive, and superior at mathematics and sciences, should become successful in their careers, and should control and suppress their feelings", whereas the woman is subjected to ambiguous and somehow contradictory demands: on the one hand, she is expected to maintain the traditional roles of nurture and care and, on the other, to take on the traditional masculine roles of success and achievement. This has contributed in shaping a new stereotype of female beauty:[4] the androgynous body. Androgyny, literally, means *male* (from the Greek *andro*) *and female* (*gyny*). The traditional body, with larger breasts and hips, would corporealize traditional roles of motherhood and nurturing. The androgynous body is flexible. It can represent masculinity – success, career, achievement and strength – and also femininity – homely, yielding and caring.

Interestingly, a female perception of beauty and a male perception of beauty do not match. A study involving a large number of male and female college students shows that males *believe* women to be attracted to men who are much bigger than what women actually like. On the other hand, the body shape that women believe is attractive to men is much thinner than what men actually find attractive.[5]

Dissatisfaction with one's own body, according to this research, is related to gender roles, and gender roles, rather than being fixed, change over time. The changes have been more salient for women, and therefore they suffer more than men from body dissatisfaction. The different types of dissatisfaction reveal and reflect the different place in society that men and women are expected to have. Men would more often than women regard a healthy body weight and shape as acceptable, whereas many women regard themselves as too fat, even if they are exceedingly slim. The dissatisfaction also covers different body areas: those males who are dissatisfied generally place the locus of dissatisfaction in their upper body (typically experienced as too small), whereas women are more often dissatisfied because they think they are 'too big', especially their lower body. Men and women diet for different reasons: for women it is a result of a pressure to be thin, and because thinness is regarded as a key attribute of female beauty. Men, instead, tend to diet for *a reason* other than that of thinness being vital for beauty or social acceptability. Reasons that induce men to diet are typically "1) to avoid childhood teasing for being overweight; 2) to improve athletic performance; 3) to avoid developing medical illnesses associated with weight problems; and 4) to improve a homosexual relationship. Unlike males, young women diet because of a common belief that weighing less is inherently good."[6]

Research on sexual orientation and eating disorders confirms the relationship between social pressure, or social values, and eating disorders. Gay men appear more prone than not only heterosexual men but also lesbian women to develop eating disorders.[7] This seems related to the fact that it is not biology that causes eating disorders, but gender roles and gender expectations.

This explains the experience of many fitness professionals who work with the general public, and who find that women suffer the most severe types of dissatisfaction, and have the most extreme behaviors.

Of course, social pressure, as I argued in Chapter 5, is not *self-explanatory*. Even if a correlation between glorification of thinness and eating disorders and other abuses of the body can be found, this still does not explain why thinness and physical strength become valuable, and why they become aesthetic requisites in the first instance, and what is attractive in thinness or sculpted bodies.

The modification of gender and sexual roles[8] can explain why women are more vulnerable than men to eating disorders and more broadly to body dissatisfaction. But a different hypothesis is necessary to explain why thinness, body sculpture and muscular tone are valued. The hypothesis, formulated in Chapter 5 (and elsewhere at greater length),[9] is that a thin and sculpted body is the emblem of body control. A skinny and super-toned body corporealizes a moral ideal, according to which *will*, *self-mastery*, *self-control* and *achievement* are reasons for moral pride. The modification of the role of the woman who has, especially in the twentieth century, been expected to demonstrate intellectual skills and to compete in areas traditionally reserved for men, can partly explain why women are more often affected by eating disorders and body dissatisfaction. It can explain why it is that *their* body is the one that is required to change – *they* have to change; and why they show the highest degree of body dissatisfaction and body abuse. However, the modification of the woman's role in society does not explain what is good about being in control, in mastering the body, and what is repugnant in fat. It is not simply femininity in the traditional sense that is rejected here. It is *the whole body* that is rejected; it is *life* that is given up, and this cannot be fully explained by social changes. This can only be understood in its profundity and tragedy by examining the conception of the good that people internalize.[10]

Conclusions

Eating disorders elicit a kaleidoscope of emotions and reactions in professionals, and people think differently about 'the right way' of dealing with the eating-disordered exerciser. Some think that accepting these participants is a part of their job, some believe it is not. Some think it is bad for clients to be turned away, and some that it is good for them not to exercise. In Chapter 9, I will explore the reasons for and against the various responses given in these case histories. I will also discuss which of these various moral strands carry more force.

It also becomes apparent that fitness professionals do not seem to have a clear idea of their legal and moral boundaries, which could assist them in determining the right course of action with people with eating disorders. None of the people I talked to have made any reference to the law or professional guidelines.[11] Whereas it is clear what the law requires of us, should someone have a heart attack in our class or should faint or have an accident (fitness professionals, at least in the UK, have to be trained in First Aid, and legislation on accident report regulates the registration of incidents – see Chapter 8), it is not clear what the law or professional codes require of us when a client with a vulnerability such as an eating disorder asks to be admitted to our clientele. The next chapter will look at both legislation and professional guidelines, and discusses whether any piece of legislation can assist us in articulating the dilemmas raised by the eating-disordered exerciser.

8 Law and professional guidelines

Introduction

Given the hazards involved in exercise for those with eating disorders, and given the growing presence of eating-disordered participants in sports and fitness environments, it is of particular importance to have some understanding of the possible legal responsibilities of those involved in the fitness industry. In this chapter I discuss some of those responsibilities, and it is important that instructors should acquaint themselves with the legal regimes applicable in their own countries and seek legal advice if a claim against them seems likely.

The principal areas of relevant law include those relating to professional negligence, contract, health and safety and – although rarely pertinent – criminal law. There is, in fact, a possible criminal liability for manslaughter if a client dies while taking supervised exercise. Much of the focus of this chapter will be on negligence, as this area of law can illuminate key concepts relating to fitness professionals' responsibilities towards their clients. These key concepts can be applied to the case of the eating-disordered exerciser. I will also discuss codes of ethical practice, which contribute to setting the standards of care which professionals can be expected to conform to.

Eating disorders raise particular issues that will be explored in this chapter, but neither legislators nor fitness professional bodies have set out clear rules for such cases. This chapter offers an overview of the principles of good conduct set up by professional guidelines, codes of ethics and law. But also, it shows that fitness professionals might be left to deal with dilemmas that are essentially ethical in nature, and to which there is no clear legal answer.

For the most part in this chapter, I confine myself to addressing the principles of common law applicable in the UK, the USA and most English-speaking countries. Civil law jurisdictions such as France and Italy also use concepts of contract and negligence law but the nature and content of the legal rules can differ markedly from common law.

Negligence in the fitness profession

One obvious question is whether the fitness professional who agrees to supervise an eating-disordered exerciser, who happens to suffer harm while training, could be liable for negligence or breach of contract. For example, might the instructor be liable if a client with bulimia has a cardiovascular collapse while exercising? Is the instructor liable for negligence if an emaciated anorexic client suffers a fracture while exercising?

In order to answer these questions, we need to look at what makes fitness professionals liable for negligence more generally.

Kevan *et al.* explain that "[t]here are very few reported cases about the liability of coaches for injuries to participants, but principle may be gleaned by reference to other areas of the law".[1]

In principle, fitness professionals could only be held responsible for a client's injury through negligence if:

1 they have a duty of care in a specific situation;
2 they have failed to meet the "standard of care that is expected and up to current standards within the industry";[2]
3 there is a causal connection between the damage or harm to the plaintiff and the defendant's careless conduct;
4 the damage is not "so unforeseeable as to be too remote".[3]

Let us clarify what these general principles mean.

Duty of care

Sometimes the existence of a duty of care can be contested.[4] That fitness professionals owe a duty of care to their supervisees, however, seems evident.

> The precise ambit of that duty will [. . .] vary greatly depending upon various factors [. . .]. The relative experience or expertise of the coach and the subject, as well as their respective ages and the nature of their relationship, will be highly relevant in determining the scope of any duty of care, whether the coach was negligent and whether there was contributory negligence.[5]

However, considering the nature of the relationship between fitness professionals and clients, it appears that, once the instructor agrees to train and supervise an applicant, and s/he becomes 'the client', the instructor owes them duty of care. The instructor is a competent, qualified individual, who agrees to instruct the participant and assist him or her with, for example, technique and exercise programming, according to his/her knowledge and qualifications. Where the client is paying for training and there is therefore a contract between the client and the professional, a duty of care will arise out of that contract. But even if a client is offered a gratuitous service (e.g. a free lesson) s/he will be relying on the expertise of the instructor and this knowledge binds the fitness professional to owe the client a duty of care. By this duty,

> [t]he coach will not instruct the participant to do any act that is likely to cause injury to the participant without warning the participant of the risk of

such injury, [. . .] he will not instruct the participant to do any act in any way that is likely to cause injury to the participant without warning the participant of the risk of such injury, [. . .] he will satisfy himself that the participant is competent to perform any act he instructs the participant to undertake.[6]

Breach of duty of care and risk

The fact that A owes a duty of care to B, and that B hurts him/herself while under A's supervision, does not necessarily mean that A is liable for negligence to B, or that B can win compensation (a sum of money for damage incurred) from A.

If, for example, the risk of the harm suffered by B is inherent to the chosen activity, A is probably not liable for negligence. The issue of whether the defendant can be held negligent for injuries that are inherent to the sport practiced has arisen in lawsuits for negligence in the context of competitive sports.[7] Many cases have been between participants, rather than between participant and trainer.[8] Sportsmen have made claims against competitors for injuries suffered on the pitch. The courts have argued that the interaction between players in competitive sports is nothing like a 'polite social interaction', and thus the same standards of care expected under the 'neighborhood principle'[9] cannot be expected among competitors.

Fitness activities are usually not competitive. Therefore they are devoid of an element of aggression towards a fellow participant. Moreover, although situations in which one participant can endanger another participant in the fitness context can be imagined (one spills water over the floor, and fails to warn others, so that a person falls and injures him/herself), we want to understand the relationship between coach – or professional more generally – and client, rather than among clients.

However, these cases highlight an issue that is pertinent to our discourse. One could ask whether, by consenting to some inherent risks, the participant takes responsibility for resulting injuries. If this were the case, the instructor would probably not be held liable for negligence for harm incurred.

It seems uncontroversial that if a client participates in a step class, s/he agrees to the minimal and rather remote risk of slipping over the edge of the step. But the risks that the eating-disordered exerciser runs while exercising are not inherent risks, but *relative* risks (see Chapter 6). They are relative to *her/his own physical state*, rather than to physical activity. Thus it cannot be assumed that the eating-disordered participant, by participating in a fitness program, consents implicitly to, let's say, fractures or heart failures.

One could of course ask whether the eating-disordered exerciser consents to the relative risks, and whether this absolves the instructor of legal responsibility. But this is a separate issue, which will be discussed later in this chapter.

For now, I have argued that the risks run by the eating-disordered exerciser while training are not inherent to the activity (differently to the risks of being harmed on the pitch during a competitive sport) and therefore we cannot presume that the sufferer consents to potential harm. From this point of view, this implies that some responsibility might fall on the fitness instructor for the harm that the client might suffer.

Breach of duty of care

The second element in determination of negligence is *breach* of the duty of care. The duty of care is quantified and qualified in terms of a *standard* that is reasonable to expect of a responsible professional. This has been established from legal healthcare cases. These cases, although specifically pertinent to clinical practice, are relevant to our discourse. The meaning of the notions and the concepts that are established in these cases, in fact, also apply to the sports and fitness context. These cases draw general conclusions as to when professionals (not only medical professionals) act in a substandard way or are responsible for harm to others (patients or clients).

One of these cases was that concerning Mr Bolam[10] (hence the term 'Bolam test'). In 1954, John Bolam was admitted to a psychiatric ward. He was diagnosed with depression. It was decided that he should be treated with ECT – electro-convulsive therapy. Nowadays, ECT is only administered for severe depression under a general anesthetic and with use of a muscle relaxant. In 1954, the use of a muscle relaxant was not as established as it is today. It was known, however, that the use of ECT without muscle relaxants might cause spasms, which could result in torn muscles and fractured bones. In the psychiatric ward to which Bolam was admitted, the policy was not to administer muscle relaxants. The doctor in charge of Mr Bolam, in accordance with hospital policy, did not administer muscle relaxants, and Mr Bolam ended up with a fractured hip. The treating doctors, however, were not held negligent toward Mr Bolam, because it was ruled that the defendants (the doctors) acted in accordance with a practice accepted as proper by a responsible body of medics skilled in that particular treatment at the time.

This case set up a legal precedent for breach of duty – the Bolam test. This means, in other words, that the standard of skill and care expected of a professional (trainer or doctor) will be that of a skilled professional specialized in that particular field. This also means that, in case of litigation, it is likely that the courts would require expert evidence, that is, that they would ask other professionals specialized in that particular field whether they would have done what the defendant did in the given circumstances. In principle, the Bolam test means that the defendant is not being careless and, hence, not negligent if skilled and competent professionals would have done the same in the given circumstances, even if harm to the patient or client has occurred, and even if a different body of opinion exists which would take a contrary view.[11]

In another case, a woman went to a jeweler to have her ear pierced. She subsequently developed an abscess, because the instruments used by the jeweler were not sterile. Although the abscess was imputable to the jeweler's instruments, the woman's claim failed because, in court, it was found that the jeweler did what any reasonable jeweler would do in the given circumstances, and had taken reasonable and responsible precautions that could be expected of him. A surgeon ought to use aseptic sterile instruments, but this could not be expected of a jeweler.[12]

As established in the *Bolam* case, the professional is not guilty of negligence if he or she acted "in accordance with a practice accepted as proper by a responsible body of medical men skilled in that particular area".[13]

A later case, *Bolitho*,[14] in 1997, tempered the norm established in the *Bolam* case. The obvious worry arising from the *Bolam* case is that a professional (any professional, and not only a doctor) would not be held negligent, insofar as expert witnesses were found who would act similarly to the defendant, regardless of whether or not the action taken was reasonable and responsible.[15]

In *Bolitho* the courts established that the professional opinion also needs to be reasonable or responsible. Even if a professional acted according to the opinion of the skilled and competent professional, s/he would still be acting substandardly if s/he acted unreasonably or irresponsibly. This in practice means that it is not enough to do what some other, or many others, would do in the given circumstances. It was stated:

> ... the court has to be satisfied that the exponents of the body of opinion relied on can demonstrate that such opinion has a logical basis. In particular, in cases involving, as they so often do, the weighing up of risks against benefits, the judge before accepting a body of opinion as being reasonable, responsible or respectable, will need to be satisfied that, in forming their views, the experts have directed their minds to the questions of comparative risks and benefits and have reached a defensible conclusion on the matter.[16]

The judge stated: "if in a rare case, it can be demonstrated that the professional opinion is not capable of withstanding logical analysis, the judge is entitled to hold that the body of opinion is not reasonable or responsible."[17] A professional (whether medic or trainer) must therefore ensure that the line of action chosen is also reasonable and responsible.

Although these cases concern doctors and their patients, the principles laid down by the courts apply to any profession, and hence to the fitness profession as well.[18] In the case of fitness professionals, professional and deontological codes assist in determining what skilled and responsible fitness professionals reasonably owe to their clients. I will discuss these standards later in the chapter. Before I do this, I examine the other element of negligence: causation.

Causation

Where causal connection is absent or cannot be proven, a breach of standards of care does not amount to negligence. The breach of duty must be the cause of harm. For example, in the health care context, in *Barnett v Chelsea and Kensington Hospital Management*,[19] three nightwatchmen went to a casualty department, complaining of abdominal pain after drinking tea. The doctor failed to examine them, advising that they see their own GP. One of the three men died soon after leaving the hospital. It was then found that the cause of death was arsenic poisoning. The case was brought to court, and the doctor was found to have breached his duty of care. However, expert evidence showed that, even if the doctor had fulfilled his duty of care, the watchman would have died anyway. Death, thus, was not imputable to a breach of duty of care, and consequently the doctor was not found negligent, although he was in breach of his duty of care.[20]

In the fitness context, some injuries to clients can obviously be caused by breach of standards of care. What if, in the studio where I am about to begin my class, there is a leak from the ceiling that has caused a pool of water on the floor, and I fail to warn my clients or to take other appropriate action, and one of my clients slips on the water and falls? If an injury can be attributed to this, this injury could be the result of my breach and thus I would probably be found negligent. In this case liability could also be claimed under the Occupiers' Liability Act 1957. This claim would lie against all occupiers, who generally would be managers and owners. Before I return to the relationship between instructors and exercisers, I shall briefly look at liability for premises.

Liability for premises

Accidents nearly always happen due to not one, but a number of reasons. Often, several facts come together to cause an accident. In the example above, the manager has not repaired the ceiling, I (the instructor) have not noticed the pool, the unlucky client was maybe looking in a different direction and slipped on the water. While instructors may be responsible for the part they played in bringing about the accident, other people might also be liable. In particular, managers/owners (occupiers) have a duty of care towards all visitors who enter the facility they manage or own.

Section 2 of the Occupiers' Liability Act 1957 states that:

(1) An occupier of premises owes the same duty, the 'common duty of care', to all his visitors, except in so far as he is free to and does extend, restrict, modify or exclude his duty to any visitor or visitors by agreement or otherwise.

(2) The common duty of care is a duty to take such care as in all the circumstances of the case is reasonable to see that the visitor will be reasonably safe in using the premises for the purposes for which he is invited or permitted by the occupier to be there.

Section 5 states:

(5) The common duty of care does not impose on an occupier any obligation to a visitor in respect of risks willingly accepted as his by the visitor (the question whether a risk was so accepted to be decided on the same principles as in other cases in which one person owes a duty of care to another).[21]

I shall soon return to the issues of risks willingly accepted (mentioned already above). It is important to note that liability under the acts and under common law is not mutually exclusive and, in fact, claims could be made both for breach of statutory duty and for negligence.

The situation, in any case, is usually not as clear as in the example of the ceiling leak, and so a determination of liability in many cases is not straightforward.

Complex cases

Imagine the following scenario. A man in his early 50s walks into your class, he tells you that he has a history of heart disease, but his cardiologist has advised moderate exercise. Indeed your class is of moderate intensity. If he has a heart attack while exercising under your supervision, it would be unclear that the accident is imputable to you, insofar as an accident of that sort can be caused by a number of factors such as severity of underlying illness, client vulnerability and disclosure of the nature and severity of the disease to fitness professionals. It is not clear that you have given this client substandard care. To prove negligence, it would first have to be proven that the activity you taught fell below the appropriate standards of ordinary skilled and responsible professionals exercising the profession, or that you have somehow failed to care appropriately for that individual client (for example, you haven't screened him properly[22]). Second, it would have to be proven that the resulting harm is due to your substandard practice, but given the complexity of causes that produce these types of injuries, causation would probably be difficult to demonstrate. A client who has a heart attack in a moderate intensity session might also have a heart attack while doing some minor decorations at home. Third, the client has voluntarily entered the studio, and as far as the client appears sufficiently aware of his medical condition and capable of understanding the risks and benefits that he would run during physical activity, he appears to have consented to take those risks (see '*Volenti non fit injuria*', later in this chapter).

The case of eating disorders is even more complicated than this. First, often the client does not disclose her/his problems; second, eating-disordered exercisers may be difficult to identify because many have normal or *quasi*-normal weight; third, it may be hard to determine the relative risks for the individual exerciser; and finally it is not clear whether the sufferer has the capacity to 'voluntarily' take the (somehow unforeseeable) risks associated with exercise under abnormal nutrition regimens.

Foreseeable harm

Another requisite for negligence to occur is the foreseeability of the consequences. For an instructor to be held negligent, the kind of harm must be reasonably foreseeable to him/her while – as the Australian case *Heil v Suncoast Fitness*[23] suggests – not too obvious either. In this case the instructor took walking sessions in a public park. One participant was hit by a bike and made a claim against the instructor. He argued that the instructor had been negligent in failing to disclose the risk of being hit by bikes on the walking path. It was decided that in a public park (a place that appeared appropriate for a walking session) it is obvious that there will be cyclists. Every adult should be aware of this without the instructor needing to say it. The instructor is not negligent, because the matter is of such an obvious nature that it can be reasonably expected that any adult would understand it by him/herself.

In the case of eating disorders, because many sufferers have normal weight and fail to disclose their condition, the risk could indeed be unforeseeable to the fitness professional. I have provided identification tools to spot those at risk. It is important to be able to detect eating disorders because even with a signed clear health screening questionnaire and clear lifestyle questionnaire it could be argued that the instructor should rely on *all* existing *evidence*, rather than only on signed questionnaires. Signed forms have only relative value in law; they are no more than one particular form of evidence. For example, in gathering informed consent, written and signed forms could provide evidence of an exchange of communication between doctors and patients: however, strictly speaking, written forms are not mandatory in law and do not provide uncontroversial evidence of valid consent. By analogy, signed health screenings can offer some degree of evidence and of protection for fitness professionals in case of litigation, but might not protect the instructor should s/he fail to give reasonable importance to evident signs of eating disorders (such as, for example, extreme emaciation, corrosion of fingers or teeth, extreme regimens of exercise, and other signs of ill health associated with eating disorders).

As Kevan *et al.* explain,

> [i]t is not inconceivable that a coach might be held liable in negligence merely for allowing a participant to take part in an activity at all where it should have been obvious to the coach that the participant was not competent to participate either because of some feature of the participant or because of the inherent danger involved in the activity in question.[24]

For example, "[c]learly a coach will be liable if he negligently instructs a gym member to lift a weight greater than the gym member could reasonably be expected to bear".[25] By analogy, it appears that an instructor might be liable for negligence if he or she allows a clearly debilitated eating-disordered client to participate in exercise that is designed for the averagely healthy person.

However, what degree of attention is required of fitness professionals in order to spot the eating-disordered exerciser and to assess risk? The 'reasonable standard' concept could be applied here: where the signs are so evident as to alert the ordinarily skilled and reasonable professional, then it could be expected that the professional should act accordingly. But we also know that eating disorders might not be evident, that sufferers might have normal weight, and that they often deny that anything is wrong. Thus, in order to identify the eating-disordered exerciser, arguably the professional must at times apply greater than ordinary skills. Moreover, even if the eating-disordered exerciser is identified, it might still be difficult to foresee the real possibility of harm, in absence of clear signs of severe disorders. We have seen in Chapter 7 that many cases of eating disorders concern people who are known to make themselves sick, or to take diuretics, or to have anomalous diet habits, but it is unclear whether they are unfit for physical activity. In addition to this, different individuals respond differently to the stresses caused by abnormal eating, so there is not a clear protocol that can be followed in order to avoid injuries. Harm, in the case of eating disorders, is thus not as clearly foreseeable or unforeseeable as it might be in other cases.

Moreover, some instructors, as we have seen in the previous chapter, worry about the harm that could result to the client if the client is barred from exercise. This raises other important issues. Should fitness professionals hold themselves legally responsible for harm that might occur should the client be advised to discontinue physical activity, or only for harm that might occur during exercise sessions? As we have seen in Chapter 7, some fitness professionals have accepted eating-disordered exercisers on the basis of what could happen to them elsewhere. Whereas this can be an ethically considerate way of thinking, fitness professionals are only legally responsible for what happens to their clients while they are exercising under their supervision, or due to the fitness advice they receive.

There is also a slightly different question: could fitness professionals be held partly responsible for harm that occurs outside exercise sessions because, for example, of the debilitation they contributed to cause? Imagine, for example, that a client walks out of the center after a fitness session and collapses. Could it be held that the professionals training the client are to some degree responsible for severe debilitation? It is difficult to give a straight answer to this question as it would depend on the circumstances of the case. It would be left to the judges to decide. In principle, a person can be held responsible for creating a source of danger,[26] even if s/he does not directly cause harm. This implies that each time there is reason to believe that a person is at serious risk it is best to address the problem, to warn the client and to attempt to protect oneself from exposing a person to that risk.

Volenti non fit injuria

Where a participant accepts the risks highlighted by the professional, in principle the professional is not liable for negligence. For example, if someone arrives late to the class, thus missing the warm-up, and the instructor warns the client that missing the warm-up increases the risk of soft-tissue injuries, and the client says that she takes the class under her own responsibility, the principle of *volenti non fit injuria* would probably apply.

Volenti non fit injuria "describes a defense to a claim in circumstances where it is shown that the claimant had consented to the breach of the duty of care which is alleged and had agreed to waive his right of action in respect thereof".[27]

This seems, at least in principle, to be pertinent to the case of eating disorders, where the client is often strongly determined to take exercise, and will often strenuously defend her/his eating and exercise habits.

However, "[t]here are at least three requirements for it to apply:

1 Agreement by the claimant to waive a claim against the defendant;[28]
2 This agreement must be voluntary, not due to compulsion by the defendant or external circumstance; and
3 The claimant should therefore have full knowledge of the nature and extent of the risk."[29]

The case of eating disorders raises the question as to whether requirements 2 and 3 can be satisfied in law. Anorexia and bulimia nervosa are classified as mental illnesses. Whereas mental illness does not necessarily entail incompetence according to English law, case law has established that anorexia might involve incompetence to make decisions relating to food and meaningful therapy.[30] Given that exercise is considered as a symptom of anorexia and bulimia (see Chapter 1) and that, according to the law, refusing food and therapy might not be competent choices *because they are symptoms* of eating disorders, it seems to follow that taking exercise might also not be considered as a fully competent choice and that the exerciser might also be regarded as unable to *voluntarily* agree to take the relative risks involved, because exercise is seen as a symptom of eating disorders. It could be argued that anorexia creates a compulsion to exercise, in the same way as it creates a compulsion to diet, regardless of the risks associated with it.[31]

Volenti in some cases might even be counter-intuitive. Imagine that a severely emaciated person walks into my step-aerobics class. Bearing *volenti* in mind, I approach her and say: 'I think you are too thin to perform this class safely. Indeed, I think you cannot even walk home safely. I need to warn you that you are in danger of fractures and cardiovascular collapse, which could result in brain damage or death. If you consent to this risk and are prepared to take full ownership for any damage that you will suffer, I let you in.' Arguably, beyond the intuitively bizarre nature of this conversation, one argument could be that the eating-disordered participant will not give valid consent, because her urge to take exercise blurs her judgment and jeopardizes her competence.

I have argued elsewhere[32] that all claims that disordered behavior is 'caused by' mental illness, or similar, do not make sense. But these arguments are not central to our present concern. What matters is how the law is likely to treat these incidents, and it would probably be difficult to employ *volenti* in defense of the instructor. *Kirkham v Chief Constable of Greater Manchester Police*[33] clarified that a person who is of an unsound mind and not fully responsible for his/her actions cannot give valid consent to a risk. This was a case of a prisoner who committed suicide in prison, while the officers responsible were aware that he was suffering clinical depression. The defense in terms of consent did not succeed, because the victim could not have 'chosen' to kill himself, given his state of mind.[34] Of course fitness instructors will not always be able to make a diagnosis of eating disorders. But in cases of serious exercise abuse and eating disorders, it would probably be argued that the disordered exerciser is in no position to consent to the risk.

Interestingly, the defense of consent failed again in another case, *Reeves v Commissioner of Police for the Metropolis*.[35] Here again a prisoner committed suicide, while he was known to be at risk of suicide. Differently from the previous case, this prisoner *had no mental illness*. The House of Lords held that *volenti* could not apply because the defendant has a "responsibility to prevent the claimant's voluntary act, whatever his state of mind".[36]

The outcome of this second case was contributory negligence. Contributory negligence refers to situations in which the damage occurs partly as the result of the fault of the claimant, and partly as the result of the fault of the defendant. In these cases, the court will make decisions considering the respective share of responsibility.[37] Given the nature of eating disorders (where competence to care for one's body is a critical issue) it is dubious as to whether either *volenti* or contributory negligence could be a successful defense. Instead, it appears less dubious that where professionals have reason to believe that a client is at serious risk due to their medical conditions, they should take actions to prevent the occurrence of harm, and this is a proper understanding of the duty of care.

Key points

- Fitness professionals could be held negligent for injuries resulting from breach of duty of care and standard of care.
- It could be difficult to prove that breach of standard of care has caused the injury.
- Some sports have inherent risks: in these cases the injured party can claim no compensation.
- The case of eating disorders raises issues of *relative risks*, not inherent risks.
- For the instructor to be held responsible for the client's injury, harm should be foreseeable.
- The risks involved in exercise for eating disorders sufferers are not always foreseeable.
- The law offers no clarification or advice as to how to handle situations in which harm is not clearly foreseeable.
- *Volenti* is not likely to apply.

Professional codes: Respect for self-determination or protection of welfare?

One key issue to ascertain whether a fitness professional could be held liable in negligence is determination of the standard of care that is expected of him/her. The courts will be likely to give high regard to any codes of practice or guidelines developed by the profession itself and so professional bodies have helped to refine the concept of 'standard of care', and it is therefore imperative to look at the guidelines that have been produced. It should be noted that guidelines are not legally binding documents. They should not be applied blindly and should always be used in the interests of the client. In the medical profession, for example, departing from medical guidelines in the best interests of patients can be a way in which a doctor might discharge his/her duty of care. Yet, "guidelines will offer some evidence of what constitutes proper treatment for a patient's condition",[38] and a similar principle applies to the fitness profession.

UK Register of Exercise Professionals

In the United Kingdom the Register of Exercise Professionals (REPs), the main professional body in exercise and fitness, issued a Code of Ethical Practice in 2008. This replaces the previous Code of Practice. Some of the key points are listed in Box 8.1. This Code defines what is good practice for professionals in the fitness industry, by reflecting some fundamental ethical values, like rights, relationships, responsibilities, and standards of care. There are four principles to the REPs Code, which I cite in full in Box 8.2.[39]

Box 8.1 Code of ethical practice, key points – REPs, UK

Background

It is important to establish, publicise and maintain standards of ethical behaviour in fitness instructing practice, and to inform and protect members of the public and customers using the services of exercise professionals. Physical activity and exercise can contribute positively to the development of individuals. It is a vehicle for physical, mental, personal, social and emotional development. Such development is enhanced if the individual is guided by an informed, thinking, aspiring and enlightened exercise professional operating within an accepted ethical framework as a professional.

The role of an exercise professional is to:

- identify and meet the needs of individuals;
- improve performance or fitness through programmes of safe, effective and enjoyable exercise;
- create an environment in which individuals are motivated to maintain participation and improve performance or fitness;
- conform to ethical standards in a number of areas – humanity, relationships, cooperation, integrity, advertising, confidentiality and personal standards.

Box 8.2 Principles in the REPs Code of Ethical Practice (emphases added)

PRINCIPLE 1: RIGHTS

Exercise professionals will be respectful of their customers and of their rights as individuals.
Compliance with this principle requires exercise professionals to maintain a standard of professional conduct appropriate to their dealings with all client groups and to responsibly demonstrate:

- respect for individual difference and diversity;
- good practice in challenging discrimination and unfairness;
- discretion in dealing with confidential client disclosure.

PRINCIPLE 2: RELATIONSHIPS

Exercise professionals will nurture healthy relationships with their customers and other health professionals.
Compliance with this principle requires exercise professionals to develop and maintain a relationship with customers based on openness, honesty, mutual trust and respect and to responsibly demonstrate:

- awareness of the requirement to place *the customer's needs as a priority and promote their welfare and best interests* first when planning an appropriate training programme;
- clarity in all forms of communication with customers, professional colleagues and medical practitioners, ensuring honesty, accuracy and cooperation when seeking agreements and avoiding misrepresentation or any conflict of interest arising between customers' and own professional obligations;
- integrity as an exercise professional and recognition of the position of trust dictated by that role, ensuring avoidance of any inappropriate behaviour in all customer relationships.

PRINCIPLE 3: PERSONAL RESPONSIBILITIES

Exercise professionals will demonstrate and promote a clean and responsible lifestyle and conduct.
Compliance with this principle requires exercise professionals to conduct proper personal behaviour at all times and to responsibly demonstrate:

- the high standards of professional conduct appropriate to their dealings with all their client groups and which reflect the particular image and expectations relevant to the role of the exercise professional working in the fitness industry;

- an understanding of their legal responsibilities and accountability when dealing with the public and awareness of the need for honesty and accuracy in substantiating their claims of authenticity when promoting their services in the public domain;
- a responsible attitude to the care and safety of client participants within the training environment and in planned activities ensuring that both are appropriate to the needs of the clients;
- an absolute duty of care to be aware of their working environment and to be able to deal with all reasonably foreseeable accidents and emergencies – and to protect themselves, their colleagues and clients.

PRINCIPLE 4: PROFESSIONAL STANDARDS

Exercise professionals will seek to adopt the highest level of professional standards in their work and the development of their career.
Compliance with this principle requires exercise professionals to commit to the attainment of appropriate qualifications and ongoing training to responsibly demonstrate:

- engagement in actively seeking to update knowledge and improve their professional skills in order to maintain a quality standard of service, reflecting on their own practice, identifying development needs and undertaking relevant development activities;
- *willingness to accept responsibility and be accountable for professional decisions or actions, welcome evaluation of their work and recognise the need when appropriate to refer to another professional specialist;*
- a personal responsibility to maintain their own effectiveness and confine themselves to practice those activities for which their training and competence is recognised by the Register.

Any alleged professional misconduct or avoidance of compliance with the terms of membership of the Register will be referred to the Professional Practice Committee which will consider any need for sanctions against an individual instructor, coach, trainer or teacher. The appropriate authority(ies) will deal with any criminal allegations.

Principle 1 ('Rights') stresses the duty of fitness professionals to respect the rights of individuals. Interestingly, the previous version[40] of this code stated, at 1(1), that every individual has a right to participate in exercise and, at 1(2), that individuals' right to self-determination should be respected. Similarly, 2(4) ('Relationships') of the previous version stated that instructors "should encourage and guide their customers to accept responsibility for their own behavior and actions in training".

These previous statements have been tempered in the revised code and included in a broader statement, requesting fitness professionals to respect "the rights of individuals": indeed, in the fitness industry, it is reasonable to interpret these rights as including the moral entitlement to participate in exercise.

The open references to respect for self-determination, instead, have been replaced by stress on respect for the 'needs' and 'best interests' of the participants. The mitigation of references to respect for self-determination is somehow in opposition with liberal principles of respect for autonomy. Other codes of ethics practice and professional standards maintain the open reference to the principle of respect for self-determination, or autonomy (see below). People are allowed, for example, to undertake risky sports and activities, such as bungee jumping and climbing. In liberal societies individual freedom is generally not restricted unless third parties ought to be protected from harm (there are exceptions to this general rule, but restrictions of people's freedom for their own good are normally kept to a minimum – see Chapter 9). From this point of view, clients should be informed of the risk and benefits of the activities they wish to take, and once this information has been provided they should be allowed to decide whether or not they wish to take that risk and their choice should be respected.

In the previous version of the REPs Code, section 2(4) also stated that instructors should "[a]lways promote the welfare and best interests of their participants". Moreover 3(9) ('Professional responsibilities') suggested that instructors, "[w]ithin the limits of their control, have a responsibility to ensure as far as possible the safety of the participants with whom they work".

It is clear that promotion of welfare and of best interests can collide with respect for autonomy, as is often the case with an eating-disordered participant. If we respect self-determination, we risk endangering health. The old code set up two ethical principles: *respect for self-determination* and *protection of health and best interests*. Where these came into conflict, the old code did not explain which of the two principles should prevail. The new code has seemingly addressed this conflict by emphasizing the *needs and best interests* of the participants, and by eliminating reference to the right to self-determination. This has, however, only nominally resolved the potential conflict between respect for self-determination and safety (more broadly, welfare).

One could object that the code has taken a clear stance: it has not simply polished off a conflict *terminologically*, that is, by removing mention of it; it has *prioritized* needs and health over self-determination. The code should be interpreted in the following sense: where conflict arises between respect for self-determination and welfare, the instructor ought to promote welfare. However, this interpretation still does not resolve the potential conflict, as the case of eating disorders illustrates: the words 'need' and 'best interests' which appear in the code also include the need for, and the interest in, *self-determination*. It might clearly be in the interests (if not in the 'best' interests) of someone with anorexia nervosa to participate in exercise. It is obviously a core need of many eating disorders sufferers to take physical activity. Thus, even if we erase the terms 'respect for autonomy' and 'self-determination' from the vocabulary of fitness responsibilities, the conflict will appear with equal strength under the umbrella of 'needs' and 'best interests'. The conflict that the new code has tried to kick out the door, has got back in through the window.

Indeed, the case histories reported in Chapter 7 show that many fitness instructors realize that exercise fulfills an important need of the eating-disordered participant and that being accepted in the fitness environment might, in many cases, be in the sufferer's overall best interests. Yet, we have seen that exercise can endanger the sufferer's health, and thus, in another sense, exercise is against their best interest: it certainly is not in their best interests to exercise if they suffer severe injury or even die, like the gymnast Christy Henrich (discussed in the Introduction).[41] In fact, as noted earlier, there is possible criminal liability for fitness professionals in such cases.

The UK Code of Practice, in conclusion, establishes important ethical principles of conduct, but does not suggest whether fitness professionals should encourage a responsible use of self-determination or protect people's welfare, and what the expected standard of care is when these two might conflict, such as in cases of eating-disordered clients.

The American Council on Exercise (ACE) Principles of Professional Conduct[42]

As an ACE-certified Professional, I am guided by the American Council on Exercise's principles of professional conduct. Whether I am working with clients, the public or other health and fitness professionals. I promise to:

- provide safe and effective instruction.
- provide equal and fair treatment to all clients.
- stay up-to-date on the latest health and fitness research and understand its practical application.
- maintain current CPR [cardiopulmonary resuscitation – my note] certification and knowledge of first-aid services.
- comply with all applicable business, employment and intellectual property laws.
- uphold and enhance public appreciation and trust for the health and fitness industry.
- maintain the confidentiality of all client information.
- refer clients to more qualified health or medical professionals when appropriate.
- establish and maintain clear professional boundaries.

ACE may take disciplinary action in the case of clause 6: "Negligent and/or intentional misconduct in professional work, including, but not limited to, physical or emotional abuse, *disregard for safety*, or the unauthorized release of confidential information" [my emphasis].

Both the REPs and the ACE guidelines stress the importance of confidentiality. I now turn to these issues of confidentiality, before commenting on codes of ethical practice. Other codes of ethics will be found in the Appendix at the end of this chapter.

Confidentiality

Fitness professionals have both a moral and legal duty to maintain confidentiality. It is important that fitness professionals acquaint themselves with the regulations that apply in their own country.

In England, in the health care context, some information may be shared within the health care team in the interests of the patient, without breach of confidentiality.[43] The patient might be considered to give implicit consent to disclosure within the team, for the purpose of provision of proper health care.[44] In the fitness context, by analogy, it could also be assumed that, in absence of evidence of the contrary, the client agrees to information sharing that is necessary to provision of adequate exercise programming and monitoring. It is in fact essential that anyone who is responsible for designing a training program or supervising a client has access to PAR-Q questionnaires and medical history.

There might be a more stringent duty to maintain confidentiality, where information is received directly from a client. Suppose, for example, that a client comes to you personally and tells you that she has bulimia. You know she also attends other classes and other instructors might be unaware of her condition. In principle you could, in the client's best interests, share information that is essential to the training of the client with your colleagues and they would be bound to the same confidentiality as you. But if the client has confided in you as a friend, in an informal way and maybe outside the fitness environment, or has asked openly to keep certain disclosures private, you are in principle bound to keep such disclosures confidential. Whereas you would be justified in disclosing information if the client suffered a medical condition that threatens others (for example, an infectious disease), it is difficult, at least in England, to justify disclosure on that basis that it is purely in the client's best interests. In these cases, the best line of action is to try to persuade the client to share the information with other members of the team, and explain clearly why this is so.

It would however be prudent to insert a clause in the contract, where the client is asked to consent to disclosure of information within the team, for the purposes of designing adequate exercise programs and supervising the client appropriately.

Comments on the codes of ethics

These ethical codes show that the fitness industry is aware of the ethical responsibilities that might fall on the professionals. Invariably, all codes stress the centrality of the welfare of the client/participant. Some, like the National Council on Strength and Fitness (NCSF) and the National Institute of Health Sciences (NIHS), also emphasize the importance of respecting the client's autonomy.

Nonetheless, in determining standards of care expected of fitness professionals, in cases in which self-determination and welfare collide, such as in the case of eating disorders, codes of ethics provide little guidance. Indeed, the notion of 'welfare' itself is open to interpretation, and for someone with eating disorders, welfare might sometimes be promoted by suspending physical activities, and sometimes by allowing the participant to take exercise in the face of relative risks.

Whereas some of these codes of ethics contain broad statements which require interpretation (for example, the ACE clause 6 on "disregard for safety" could be a basis for disciplinary action, yet what defines a disregard of safety is not specified), other codes of ethics contain important core concepts which could direct decisions in the management of the eating-disordered exerciser.

The first important concept is in the Fitness Standard Council Code, at point 15, and in the NCSF code (at II.1 and IV.2). These clauses say that fitness professionals should not attempt to undertake tasks that are beyond their qualifications and expertise. I shall return to this point at the end of Chapter 9, and see how this applies to the ethical reasoning concerning the eating-disordered exerciser.

The second important core concept is that fitness professionals should liaise with other professionals and consult health care professionals any time they come across a situation that is medically complex and beyond their competence and expertise (NCSF IV.2; NIHS II.3). This implies that a network of qualified personnel should rotate around the sports and fitness arena, and that fitness professionals should have clear routes to enable them to open communication with other health specialists. This is often in practice not the case, and whereas in the UK GP referral schemes are now beginning to operate,[45] in the vast majority of cases health care professionals are not available to fitness professionals for consultation, and fitness professionals have to make decisions that involve judgment over medical matters which they are not qualified to make.

The third important concept is that it is *unethical* to either prescribe a program that is of no benefit to the client (NCSF I.6; NIHS, I.7), or to prescribe a program that over-trains the client (NIHS, I.10). This point is particularly relevant to the case of eating disorders, where there is a fine line between *useless training* and *over-training*. For example, in Case history 1 in Chapter 7, M. talked about *non-training training* for eating-disordered exercisers: a normal program of training would amount to over-training. One question is whether it is ethical to *pretend* to train a client. Insofar as *benefit*, like *welfare*, is a wide-ranging concept, which includes psychological satisfaction, a program of fitness that does not overload the client seems preferable to any other option. As a general rule, however, if a client is unfit to exercise, she should be asked to suspend physical activity. Telling white lies to satisfy customers is a form of paternalism that is not compatible with true respect.

The final important concept that some codes stress is *the social role* of the fitness profession. This can be implemented through the education and establishment of a relationship of mutual trust and respect with clients. In order to do this, however, fitness professionals should possess relevant information on eating disorders. It would be essential that, even at basic levels of professional education, information on eating disorders and exercise be delivered. I will return to this in the final chapter.

What also emerges from the codes of practice is that at the heart of every decision there should be a sound assessment of risk. Assessment is part of the standard of care expected of fitness professionals and it is to this that I now turn.

Key points

- Codes of ethics set up important ethical principles.
- Codes of ethics stress the importance of respect for self-determination.
- Codes of ethics stress the importance of protection and promotion of the exerciser's welfare.
- When these two come into conflict, as in the case of eating disorders, the codes of ethics do not clarify which of these two principles should take precedence, or how the conflict should be resolved.
- The notion of welfare is ambiguous: it could be in someone's interests to be allowed to exercise in spite of relative risks, or to be asked to suspend physical activities.
- Ethical codes stress the importance of not undertaking tasks that go beyond the professional's competencies.
- Ethical codes advise professionals to liaise with health care professionals as appropriate.
- Ethical codes stress the importance of tailoring adequate programs that will benefit the client.
- Ethical codes emphasize the important social role of the fitness profession.

Assessment

Risk assessment is one of the responsibilities of fitness instructors. Any proper understanding of the concept of 'standard of care' must include assessment, which involves identification of those at risk. Identification of people at risk of having eating disorders might be an extraordinarily difficult exercise, bound to remain unsuccessful on an indefinite number of occasions. Of course it cannot be expected of any of us that we know what is impossible to know. But it can be expected that fitness professionals do their best to gather relevant information, and that they make good use of it. How 'their best' should be understood, and what 'a good use' of information is, is not altogether clear. In what way should fitness professionals gather information? What degree of control should professionals have over their clients? How should one handle the situation with those at risk? The American College of Sports Medicine (ACSM) has provided important clarifications on the expected standards of care in assessment.

ACSM on assessment

In some countries the law requires that prior to the initiation of any supervised physical activity, clients provide medical certification. In many US states as well as in the UK, obtaining a medical certificate of clients is not a legal requirement.

The American College of Sports Medicine (2006)[46] recommends that all fitness and sports facilities and any facility offering physical activities and services should conduct a *health and fitness screening* of prospective users or new members ('Physical Activity Readiness' Questionnaires, or PAR-Q). At the end of the chapter two samples of PAR-Qs can be found.

PAR-Q forms

> should be given out to all individuals and completed prior to the start of an exercise program. The form should then be interpreted by a suitably qualified professional and dealt with in a confidential manner. If the instructor feels that the form reveals anything that is beyond the capacity of their training or experience, they should refer the case to a suitably qualified colleague or supervisor.[47]

Where there is no better qualified member of staff, it is unclear what an instructor should do. More advice will be offered in Chapter 9 and in the Conclusion.

Possible supplements to the PAR-Q are Physical Activity Screening Questionnaires and Lifestyles Questionnaires.

The *Physical Activity Screening Questionnaires* ask questions about the type of physical activity, if any, performed by the applicant, together with the intensity and duration. They might also ask questions such as: "Can you jog for 20 minutes without fatigue?"; or "Can you walk briskly for 30 minutes without fatigue?"; "If you are not currently exercising, when was the last time you participated in regular exercise?"[48]

Lifestyle Questionnaires ask questions about smoking, alcohol consumption, diet, working patterns (for example, how many hours per day one spends at a desk).

The ACSM provides specific and detailed information relating to how the screening should be conducted and what steps should be taken by fitness professionals, following collection of these questionnaires. The ACSM also recommends lines of action, following the analysis of data obtained through health screening. More particularly, it draws a table of risk factors (known as stratification of risk), which includes, for example, family history of coronary heart diseases (CHD), smoking, hypertension, sedentary lifestyle and others.[49]

Starting out from this table, it suggests that men of less than 45 years of age and women of less than 55 years of age are to be included in the *low risk category* if they meet no more than one of these risk factors. Thus, a 50-year-old woman who does not take the recommended 30 minutes of moderate activity at least three times a week is in the low risk category. If, for the same age group, people meet two or more of these risk factors they are in the *moderate risk category*. People above 45 (males) and 55 (females) are also in the *moderate risk category*. The *high risk category* includes those with known cardiovascular, pulmonary or metabolic diseases. Chronic pulmonary disease, or Type 1 or 2 diabetes, invariably indicate high risk of CHD.

In general, whether or not exercise is contraindicated, and whether or not the physician's consent is recommended, depends on the category of risk where one falls, and on the type of exercise chosen. Those in the *high risk category* should either obtain medical certification or be supervised by specially qualified personnel. Those in the *moderate risk category* who wish to undertake low to moderate physical activities (up to 60 percent of VO2Max[50]) should typically be accepted without a medical clearance form. However, those in this category who want to take strenuous exercise should be encouraged to provide a medical certificate.

Discretion, thus, ultimately falls on the instructor. S/he has to identify risk categories for each client, and verify whether the desired exercise is low, moderate or high intensity for that particular client. This is not necessarily an easy task, given that intensity of exercise is subjective, and each individual, depending on a number of factors, such as body weight, level of fitness, age and experience, fatigues differently.

The problem of assessment of intensity is sometimes resolved through fitness testing. However, fitness testing should be avoided if the client appears debilitated. The American College of Cardiology (ACC)[51] has drawn up a list of conditions that represent absolute contraindication to exercise testing. Among these are recent myocardial infarction, unstable angina, acute pulmonary embolus or pulmonary infarction, and others.[52] It has also provided a list of conditions that represent *relative contraindications*. Among these are electrolyte abnormalities (i.e. hypokalemia, hypomagnesemia), tachydysrhythmia or bradydysrhythmia, neuromuscular, musculoskeletal or rheumatoid disorders that are exacerbated by exercise, and mental or physical impairment leading to inability to exercise adequately. For these, the ACC recommends that testing be avoided, unless the benefits of exercise testing supersede their risks.

These are all relevant to the eating-disordered client who typically suffers some of these medical conditions. The eating-disordered client might thus be unsuitable for exercise testing, and thus monitoring intensity of exercise for her could be difficult. In spite of the specificity of these guidelines, especially those relating to stratification of risk, ultimately it is left to the professional's discretion to act according to knowledge and experience.

I shall now see what standard of care can be expected when an assessment of the type recommended by the ACSM is unattainable, such as in situations of group exercise.

> **Key points**
>
> - Assessment of risk is recognized by the ACSM as one of the fitness professional's responsibilities.
> - There is no legal requirement to perform such an assessment, but the ACSM recommends a standard of care which is accepted in the UK as well as in the US.
> - In case of litigation, whether there is evidence of assessment the professional view will be important.
> - The ACSM offers recommendations as to when physical activity should be interrupted or allowed only after medical certification. However, the case of eating disorders cannot easily be put in the categories of risk identified by ACSM.

What should fitness professionals do when PAR-Qs are unattainable?

In some areas of fitness, screening of the kind advised by ACSM, and a subsequent stratification of risk, is impossible. In the context of group exercise, for example, it is unrealistic to expect that group instructors obtain PAR-Qs from each participant. Group instructors who work in sports centers do not have an obligation to obtain screening questionnaires of their clients. It is understood that the managers should obtain them, and allow clients in the studio only after they have completed their screening. However, it is also common practice for many sports centers to offer 'drop-in' classes. This is consistent with other areas of sports and recreation: we might access many swimming pools without undertaking any health screening; we might join organized bike or trekking tours without prior health screening. Many group instructors will have both screened regular members and unscreened 'drop-in' participants, and often they do not know who has completed the screening and who has not. It would be unrealistic to expect group instructors to obtain a signed PAR-Q from every casual participant, not only because this might use up the time of all those others who have come for exercise, but also because, at least in England, this is not ordinary practice, and fitness professionals, like any other professional, are expected to conform to the accepted standard of care accepted at a particular time (see above, 'Breach of duty of care'). In the context of sports and fitness centers, group exercise instructors often choose to address verbally the issue of whether there are medical conditions people would like to tell them.[53]

It can be deduced by the analysis of pertinent legislation and recommendations that the standard of care expected of a fitness professional includes:

1 Risk assessment appropriate to role, qualification and area of expertise;
2 Supervision of every client accepted (this applies equally to group exercise);

3   Explanation of correct technique of exercise and use of equipment;
4   Design of activities as appropriate to the health, age and fitness level of participants (individual participants, if they are involved in one-to-one sessions, or participants as a group, if they are group instructors);
5   Provision of specific training programs for clients with medical conditions or diseases (in the context of group exercise, offer of appropriate alternatives);
6   Identification of any potential hazard in the fitness/exercise environment;
7   Request of discontinuation of exercise to those reasonably considered at high risk.

In the context of management of eating disorders, this implies that, with or without signed clear health screening, a suspicion that a client has an eating disorder should lead the professional to take *some steps* to prevent harm to the client. Chapter 9 will discuss the ethical avenues that could be taken in this case, and what these steps might be. A signed PAR-Q is not sufficient to prove suitability to engage in fitness activities, when there is clear evidence of ill health about which the client has given no information. If evidence shows that a person is at serious risk, exercise should be modified accordingly and, in the most severe cases, a request of suspension of activity is within the legitimate powers of fitness professionals. Indeed, not requesting suspension of activity where risk seems serious and imminent could expose the professional to claims in negligence, whether or not a PAR-Q has been signed, and whether or not the client has given open consent to take the risks associated with exercise while under the stress of eating disorders.

Conclusions

Fitness professionals owe a duty of care to their clients, according to the standards which are shaped by professional bodies through codes of ethical practice and professional guidelines. However, these do not address all doubts, and genuine ethical dilemmas might remain. Fitness professionals encounter a variety of clients who approach the world of fitness with a kaleidoscope of histories, medical conditions, psychological and physiological needs and hopes. Understanding whether and in what ways physical activity can benefit or harm them can be an incredibly difficult exercise. Eating disorders, in this already complex panorama, represent a unique challenge, due to the hidden nature of the condition, and to the fact that the goals of the eating disorders sufferer are sometimes at odds with the appropriate goals of the fitness profession. In severe and evident cases it is clear that by law suspension of activities must be requested. Allowing a participant to risk his or her life to participate in an exercise session is arguably a breach of duty of care. But in less extreme situations, or at least in *apparently* less extreme situations, significant discretion is left to the professional, who is called to make best use of his competencies and reasoned judgment to make decisions on specific instances. This discretion, a precious resource in any human relationship, can only be exercised if professionals have sufficient understanding of the benefits and risks of exercise for eating disorders sufferers, and of the legal and ethical responsibilities that they have.

The codes of ethics emphasize the importance of shared values of respect, protection of health, respect for boundaries and social role of physical educators. Yet, the conflict between respect for self-determination and protection of welfare, which is at the heart of the relationship between almost every health professional and the eating disorders sufferer, is not addressed in a fully satisfactory way. The guidelines leave discretion to the professionals, and might not in practice help the many who, on a daily basis, have to make decisions relating to whether, and, if so, how they should help an eating disorders sufferer to exercise.

Screening an eating-disordered exerciser could also be extremely unproductive: self-report of eating disorders sufferers is notoriously unreliable.[54] Thus, if health screenings may provide some legal protection in case of injury, they might have little value in terms of tailoring appropriate exercise and in terms of sports and fitness ethics. In other words, they may help fitness professionals in the unfortunate event of litigation, but will not always resolve the moral dilemmas of the instructor.

The next chapter will propose an analysis of the ethics of managing people with eating disorders in the exercise context. There are different ways of thinking about the ethics concerning the eating-disordered exerciser. I will discuss the strengths and weaknesses of these various ethical avenues.

Physical Activity Readiness Questionnaire (PAR-Q): Sample 1

Please check YES or NO.

YES   NO

❏   ❏   Has your doctor ever said that you have a heart condition *and* that you should only do physical activity recommended by a doctor?

❏   ❏   Do you feel pain in your chest when you engage in physical activity?

❏   ❏   In the past month, have you had chest pain when you do physical activity?

❏   ❏   Do you lose balance because of dizziness or do you ever lose consciousness?

❏   ❏   Is your doctor currently prescribing drugs (for example, water pills) for your blood pressure or heart condition?

❏   ❏   Do you know of any other reason why you should not do physical activity?

If you answered YES to one or more questions:
Talk with your doctor by phone or in person BEFORE you start becoming much more physically active. Tell your doctor about this PAR-Q questionnaire and get their consent to become more active.

I have read, understood, and completed this questionnaire. Any questions that I had were answered to my full satisfaction.

Name: _____   Date: _____

Signature: _____

Witness
This form must be witnessed at the time of signing and the witness must be 18 or over.

Name: _____

Signature: _____

Physical Activity Readiness Questionnaire (PAR-Q): Sample 2

Please read these questions carefully and check the yes or no opposite the question if it applies to you.

Top of Form

	YES	NO	
1.	❏	❏	Has your doctor ever said you have heart trouble?
2.	❏	❏	Do you frequently have pains in your heart and chest?
3.	❏	❏	Do you often feel faint or have spells of severe dizziness?
4.	❏	❏	Has a doctor ever said your blood pressure was too high?
5.	❏	❏	Has your doctor ever told you that you have a bone or joint problem such as arthritis that has been aggravated by exercise, or might be made worse with exercise?
6.	❏	❏	Is there a good physical reason not mentioned here why you should not follow an activity program even if you wanted to?
7.	❏	❏	Are you over 65 and not accustomed to vigorous exercise?

Bottom of Form

- **If you answered YES to one or more questions . . .**
  If you have not recently done so, consult with your personal physician by telephone or in person before increasing your physical activity and/or taking a fitness test.

- **If you answered NO to all questions . . .**
  If you answered the PAR-Q accurately, you have reasonable assurance of your present suitability for an exercise test.

Appendix: Standards of care and ethical codes

THE US Fitness Standards Council:[55] Code of Ethics for Personal Trainers [emphases added]

As a member in good standing of the Fitness Standards Council, I will do my utmost to:

1 Provide quality instruction to all clients and customers.
2 Screen health, exercise and nutrition history by having all clients complete a standard questionnaire form and, if necessary, complete a pre-exercise testing and/or doctor's physical.
3 Provide modified training programs for clients with special requirements or fitness levels.
4 Obtain specialized education for training special groups, e.g., the mentally challenged, osteoarthritis, etc., if applicable to my business operations and services offered.
5 Investigate, learn and incorporate valid, current research in the programs created.
6 Apply logic and objectivity, not emotion and subjectivity to all procedures.
7 Be knowledgeable and competent in CPR and emergency procedures while providing and maintaining a safe and effective training environment.
8 Encourage commitment to quality nutrition and fitness as a lifelong goal by following a regular exercise program and proper dietary habits.
9 *Educate clients about the benefits of a healthy lifestyle and promote self-reliance when appropriate.*
10 Track the progress of clients and offer assistance when needed.
11 *Respect the confidentiality of my clientele, never to disclose to any person any personal information acquired by a client unless authorized, in writing, by the client, unless a risk of health or death becomes an issue.*
12 Avoid sensual or sexual touching and maintain professional touching in cases of technical spotting, correcting technique or posture, or when conducting necessary assessments.
13 Uphold a professional image by dressing appropriately, refraining from unhealthy practices and maintaining a positive attitude.

14 Accurately represent my qualifications, avoid misleading advertising, and accept responsibility in all my actions and fitness recommendations.
15 *Recommend other professionals in fields outside my expertise.*
16 Respect the rights, welfare, and dignity of fellow professionals and exchange knowledge and experience for the benefit of the exercise and nutrition science industries.
17 Never discredit other fitness practitioners, facilities, or organizations publicly through comment or conduct unless an objective opinion serves to educate the public about low quality or unsafe or dangerous practices.
18 Conduct business in a manner that commands the respect of the public within this industry and for the goals in which we strive.
   [. . .]
25 Never discriminate on the basis of sex, race, age or *mental and/or physical limitations, and strive to provide equal and fair treatment to all individuals.*
26 *Expose irrationality that promotes potential danger or injury to individuals who do not have the expertise or knowledge to know better,* and to safeguard the public by reporting violations to the FSC.
27 Comply with local, state/provincial and federal laws of the land including, but not limited to, applicable business, medical, employment, and copyright laws.

The National Council on Strength and Fitness (NCSF): Code of Ethics[56] [emphases added]

NCSF Certified Personal Trainers – as members of the Health profession, have ethical responsibilities to their clients, society, *as well as to other health professionals.* The following ethical foundations for professional activities in the field of personal training and health promotion serve as a Code of Conduct for practicing professionals. The Code implements many of these foundations in the form of rules of ethical conduct. [. . .]

Ethical Foundations

I. The Trainer–Client relationship: *The welfare of the client is central to all considerations in the trainer–client relationship.* Included in this relationship is the *obligation of the trainer to respect the rights of clients,* colleagues, and other health professionals. The respect for the right of individual patients *to make their own choices about their health care (autonomy) is fundamental.* The principle of justice requires strict avoidance of discrimination on the basis of race, color, religion, national origin, or any other basis that would constitute illegal discrimination (justice).

   [. . .]

IV. Professional relations: The personal trainer should respect and cooperate with other personal trainers, fitness instructors and allied health professionals.

V. Societal responsibilities: The personal trainer has a continuing responsibility to society as a whole and should support and participate in activities that enhance the community. As a member of society, the personal trainer must respect the laws of that society. As professionals and representatives of the NCSF, personal trainers are required to uphold the dignity and honor of the profession.

Code of Conduct

I. Client–Trainer Relationship

1 *The Client–Trainer relationship is the central focus of all ethical concerns, and the welfare of the client should form the basis of all medical judgments.*

2 The Trainer should serve the clients by exercising all reasonable means to ensure *that the most appropriate training and fitness recommendations are provided to the client.*

3 The Client–Trainer relationship has an ethical basis and is built on confidentiality, trust, and honesty. The trainer must adhere to all applicable legal or contractual constraints while in the client–trainer relationship.

4 Sexual misconduct on the part of the trainer is an abuse of professional power and a violation of client trust. Sexual contact or a romantic relationship between a trainer and a current client is always unethical.

5 The trainer has an obligation to obtain the informed consent of each client. In obtaining informed consent for any course of physical measurement or activity, the trainer should present to the client, or to the person legally responsible for the client, in understandable terms, pertinent facts and recommendations consistent with good professional practice. Such information should include alternate modes of testing or physical activity and the objectives, risks, benefits, possible complications, and anticipated results of such activities or testing protocols.

6 *It is unethical to prescribe, provide, or seek compensation for therapies or products that are of no benefit to the client.*

7 The trainer should respect the rights of clients, colleagues, and others and safeguard client information and confidences within the limits of the law. If during the process of providing information for consent it is known that results of a particular test or other information must be given to governmental authorities or other third parties, that should be explained to the client.

8 The trainer should not discriminate against clients based on race, color, national origin, religion, or on any other basis that would constitute illegal discrimination.

II. Trainer Conduct and Practice

1 *The trainer should recognize the boundaries of his or her particular competencies and expertise, and provide only those services and use only those techniques for which he or she is qualified by education, training, or experience.*

[. . .]
3   In any training environment, the trainer should exercise careful judgment and take appropriate precautions to protect the client's welfare with regards to equipment, facilities and environmental factors.
[. . .]

IV. Professional Relations

1   The trainer's relationships with other trainers, fitness directors, physicians, physical therapists, and other health-care professionals should reflect fairness, honesty, and integrity, sharing a mutual respect and concern for the client.
2   *The trainer should consult, refer, or cooperate with other trainers, health professionals, and institutions to the extent necessary to serve the best interests of their clients.*
[. . .]

V. Societal Responsibilities

1   The trainer should support and participate in those health programs, practices, and activities that contribute positively, in a meaningful and effective way, to the welfare of individual clients, the health fitness community, or the public good [. . .]

National Institute of Health Sciences (NIHS): Code of Ethics [emphases added]

You, as a member of the health profession, have an ethical and moral responsibility to your client, community, and other health professionals in all health and wellness related services.

The following ethical foundations for professional activities in the field of personal fitness training and health promotion serve as a **Code of Conduct** for practicing professionals. The Code implements many of these foundations in the form of rules of ethical conduct. Noncompliance with the Code may affect an individual's initial or continuing status as a recognized certified professional in the National Institute of Health Science and other certifications related to their field.

Ethical Foundations

I. **The Trainer–Client relationship**: The welfare and safety of the client is central to all considerations in the trainer–client relationship. Included in this relationship is the obligation of the trainer to respect the rights of clients, colleagues, and other health professionals. *The respect for the right of individual clients to make their own choices about their health care (autonomy) is fundamental*. The principle of justice requires strict avoidance of discrimination on the basis of race, color, religion, national origin, or any other basis that would constitute illegal discrimination (justice).

[. . .]

IV. **Professional relations**: The personal fitness trainer should respect and cooperate with other personal fitness trainers, fitness instructors and allied health professionals regardless of their certification, affiliation or profession. This respect and cooperation does not supersede any obligations that the NIHS certified fitness professional has to protect the safety and welfare of the client.

V. **Community responsibilities**: The personal fitness trainer has a continuing responsibility to the community as a whole and should support and participate in activities that enhance the community. As a member of society, the personal trainer must respect the laws of their state. As professionals and representatives of the NIHS, personal fitness trainers are required to uphold the dignity and honor of the profession.

VI. **Client Confidentiality**: The personal fitness trainer has the responsibility of securing all information of a personal nature, including health status, training programs, and measurements or assessments. Client information must be placed in a secure location only accessible to the trainer or other certified trainers. Fitness trainers must consider the privacy of their clients before discussing client information with other health professionals. It is unethical to discuss such information in a public forum or in an environment in which others may gain access to that client information. Information may not be released to other entities, regardless of their intention without the written authorization of the client in the form of a release. It is further considered the right of the client to access and obtain copies of their records, including workouts and other information contained within their file.

[. . .]

VIII. **Medical Advice**: A trainer is not permitted to give medical advice, including the diagnosis of health related conditions, their prognosis or recommendations for the treatment of such medical conditions. This is including but not limited to physical therapy, acute stage rehabilitation or the recommendation of any drugs, including over-the-counter medication.

Personal Code of Conduct

I. Client–Trainer Relationship

1   *The Client–Trainer relationship is the central focus of all ethical concerns, and the welfare of the client should form the basis of all judgments for program design and assessments.*

2 The trainer should serve the client by exercising all reasonable means to *ensure that the most appropriate training and fitness recommendations are provided to the client.*

3 The Client–Trainer relationship has an ethical basis and is built on confidentiality, trust, and honesty. The trainer must adhere to all applicable legal or contractual constraints while in the Client–Trainer relationship.
   [. . .]

5 *The trainer has an obligation to obtain the informed consent of each client.* In obtaining informed consent for any course of physical measurement or activity, the trainer should present to the client, or to the person legally responsible for the client, in understandable terms, pertinent facts and recommendations consistent with good professional practice. Such information should include alternate modes of testing or physical activity and the objectives, risks, benefits, possible complications, and anticipated results of such activities or testing protocols.

6 *Exercise professionals must practice a standard of care that includes basic pre-screen or exercise readiness forms such as a PAR-Q or other risk factor analysis on all clients.*

7 *It is unethical to prescribe, provide, or seek compensation for therapies or products that are of no benefit to the client.*

8 The trainer should respect the rights of clients, colleagues, and others and safeguard client information and confidences *within the limits of the law.* If during the process of providing information for consent it is known that results of a particular test or other information must be given to governmental authorities or other third parties, that information should be explained to the client.

9 The trainer should not discriminate against clients based on race, color, national origin, religion, or on any other basis that would constitute illegal discrimination.

10 *The trainer has an obligation to their client to provide services that have been agreed to by the client and the trainer. It is a violation of professional conduct to participate in practices that inconvenience a client including late session starts, over-training of the client and no shows for sessions.*
   [. . .]

II. Trainer Conduct and Practice

1 *The trainer should recognize the boundaries of his or her particular competencies and expertise, and provide only those services and use only those techniques for which he or she is qualified by education, training, or experience to provide.* A trainer may take blood pressure measurements to determine exercise readiness or eliminate risk.

2 *It is the responsibility of the trainer to ensure that their client exercises or participates in exercises that are appropriate to that client's level of fitness.*

3 *Should the trainer be qualified to work with physician-referred clients, then they must follow physician recommendations. It is unethical and a violation of scope of*

practice to recommend programs, movement patterns or resistances not approved in writing by the physician.

[. . .]

5 *In any training environment, the trainer should exercise careful judgment and take appropriate precautions to protect the client's welfare with regards to equipment, facilities and environmental factors.*

[. . .]

10 In any testing or assessments, the trainer must follow an appropriate standard of care to include appropriate screening of a client, ensuring physician approvals or referring individuals to allied professionals when appropriate to do so.

[. . .]

IV. Professional Relations

[. . .]

2 *The trainer should consult, refer, or cooperate with other trainers, health professionals, and institutions to the extent necessary to serve the best interests of their clients.*

[. . .]

V. Community responsibilities

1 The trainer should support and participate in those health programs, practices, and activities that contribute positively, in a meaningful and effective way, to the welfare of individual clients, the health fitness community, or the public good.

9 Ethical issues

This is the narration of a person with eating disorders.

"You've ruined everything. You gave in. You're weak," I whispered fiercely. The eyes in the mirror filled with tears. [. . .] My eyes fell on the red door to the handicapped stall of the stark bathroom. I walked slowly toward it, wiping my eyes on my sleeve. I took a fateful step into that stall, and tumbled down the rabbit hole.

I shut the door and slid the lock into place, oblivious to the metamorphosis that had just occurred. I looked cautiously at the white porcelain toilet with its silver handle and pushed the sleeves of my brown and cream striped shirt up to my elbows. Lifting the seat, I took a deep breath. I opened my mouth as wide as I could and slid my right index finger down my throat. [. . .]

The door creaked. I froze, terrified that I would be caught. [. . .] My heart pounded as I listened to the intruder enter the stall next to mine. I listened, petrified, as she flushed the toilet and unlocked the door. I heard the water in the sink begin to run, the hand dryer start, and finally the creak of the door signaling her exit. I turned around and thrust my finger back into my epiglottis. [. . .]

I felt clean. Empty. I had regained control. [. . .]

That Friday night, I crossed a line. My New Year's Resolution ceased to be a diet and became a disease. It progressed rapidly. I cut my caloric intake to a maximum of 1,000 calories a day, and vomited more with each passing week. Soon, I was vomiting daily, usually after dinner. I felt weak and was plagued by headaches. I didn't care. I was losing weight.

I categorized food into "safe" and "unsafe" groups. Some of the groupings were logical (candy is bad, fruit is okay), but others were completely arbitrary. Great Harvest Bread fell into the safe bracket, despite the fact that it is fairly fattening. I stopped eating meat even though some types of lean meat are healthier than processed carbohydrates. (Meat was also harder for me to throw up than foods like pasta.) I refused to drink milk, juice, or regular soda because I was convinced that liquids with calories were a waste. I lived on bread, cereal (never in bowls, just by the handful or perhaps in a plastic baggie), fat-free frozen yogurt, and fruit. Everything else wound up in the toilet. [. . .] I was constantly tired, but could not sleep at night. My hair pulled away from my scalp as I washed it in the morning. I bruised easily, and felt cold all the time. Headaches tormented me daily. Standing up too quickly left me dizzy, and my pulse plodded along stubbornly.

Worse than the physical pain, however, was the emotional and mental anguish. I could not concentrate since I thought incessantly of food. During class, instead of listening to lectures or taking notes, I thought about what I had eaten that day, when I would eat again, what I would eat, and whether I would have the opportunity to throw up. [. . .] At the beginning of the disease, I weighed myself each morning, then each morning and each night, then several times in between, until I literally weighed myself a half dozen to a dozen times a day. I thought of nothing but how I needed to be thinner. [. . .][1]

Introduction

Stories like this illuminate the drama experienced by those who are in the grip of an eating disorder. They also raise a number of ethical and clinical issues. How should we think about this condition? How can we help sufferers? What sorts of limits to their freedom of action are good for them? And what limits are ethically justified?

This book has shown that eating disorders raise acute ethical issues for sports and fitness professionals. As we have seen in previous chapters, many studies have analyzed the relationship between sports, exercise and eating disorders. Other studies have explored the ethical and legal quandaries that eating disorders pose for health care professionals.[2] In contrast, the ethical and legal issues that eating disorders raise in the sports and fitness environment have seldom been discussed. Yet, the way eating disorders challenge daily decisions in the context of physical activity is an interesting and compelling example of how the classic dilemmas of applied ethics are increasingly found outside the boundaries of health care settings.

The eating disorders sufferer is typically intelligent, skilled and bright. Many eating disorders sufferers are successful at school and in their career. There are online lists of 'celebrities' with eating disorders.[3] This illustrates how many, if not most, of those who have eating disorders are far from the stereotype of the 'mentally ill' person detached from reality and in need of continual protection and supervision. In addition to this, we have seen in previous chapters that there is no evidence that eating disorders are caused by factors that are beyond the person's willful control over their own behavior.

Can it be right to interfere with the freedom of intelligent people for their own good, when they do not cause harm to others, but only to themselves? It might appear somehow 'obvious' to some that we should protect the eating-disordered person from extreme self-harm. A person with severe anorexia for example can be coercively hospitalized and fed (at least in England and Wales, but also in many other jurisdictions). The behavior of the eating disorders sufferer seems so 'irrational' that protection of their welfare might appear a moral imperative.

However, the problematic character of the restriction of their freedom appears more clearly when one considers similar situations, in which, instead, we take it for granted that people should be free to harm themselves in similarly 'irrational' ways. Obese people cannot be forced to diet, smokers cannot be forced to quit. Why, then, should we be able to stop a person from exercising or dieting?

The philosopher John Coggon points out that there are two separate questions here: one is what mandates an interference with freedom of action; the second is whether it is ethical to be inconsistent or hypocritical. Assuming that inconsistency should be justified by relevant differences, the question is whether it is ethical to restrict eating-disordered people's freedom of action. If it is ethical, then it should also be ethical in other cases (obesity for example) unless relevant differences between the cases are shown, or unless other compelling reasons of, for example, social utility, or desirability, can persuade us that the world is on the whole a better place (or that life is more interesting, or more desirable) if people are normally allowed to harm themselves. In this chapter, and in this book more generally, I do not pursue this line of argument – namely, what would happen to our society and to all and each of us if one had to apply consistently a restriction of freedom to self-harm. I instead focus on whether the behavior of the eating-disordered person should be restricted on its own merit, and regardless of the wider social implications of this, because my conclusions anyway do not gesture towards a general restriction of freedom to harm ourselves, which could have important social repercussions.

In this chapter, I identify, discuss and propose a way of understanding and resolving the ethical dilemmas caused by the increasing presence of eating-disordered exercisers in sports and fitness settings. It should be noted that the rise in obesity rates also causes ethical problems for the fitness and sports instructors. However, there are crucial differences between obesity and eating disorders, from an ethical point of view. Some of these have already been discussed in previous chapters, but it is worth reminding ourselves here why eating disorders, rather than obesity, generate the most acute problems in sports and fitness.

First, obesity is easily identifiable. Thus the issues about identification, and about how far we should go in scrutinizing the client's private life, are not as sharp in the management of the client affected by obesity.

Second, people affected by obesity normally undertake physical activities in order to improve their health; the eating-disordered exerciser, instead, often exercises to the detriment of their health.

Third, the obese person who enrolls onto a fitness program generally knows s/he has a problem and is probably trying to deal with it. The eating disorders sufferer instead, typically either fails to perceive that s/he has a problem at all (*denial*),[4] or s/he is aware that s/he has a problem, but does not disclose it to the relevant people, and, when asked, will deny that anything is wrong. Sometimes this process is called *negation*[5] and is differentiated from denial in that denial is a form of self-deception, whereas negation is an attempt to deceive others.

Fourth, establishment of meaningful communication with the eating disorders sufferer can be extremely difficult (see Chapter 1). This problem is not so acute in the management of the obese client (although there might be problems of adherence to diet and exercise).

Eating disorders thus represent a unique challenge for the fitness professional. What will be said in this chapter will assist fitness professionals in understanding and resolving the ethical predicaments of eating disorders, but what will be argued here can be applied by extension to all situations in which clients wish to set up goals and methods of training that are different to those held appropriate by the fitness instructor. The avenues explored here can also be applied to non-professional contexts, as they provide ways of thinking about the quandaries that can arise in the sports and fitness arena, for any participant, and not just for the qualified instructor.

Allowing people with eating disorders to take exercise: Reasons for and against

Faced with an exerciser who wishes to continue to exercise in the face of advanced emaciation, or the perils of eating disorders, intuition seems to pull in different directions (see Chapter 7). This is because exercise can harm the client, but can also benefit them and help them to achieve better body satisfaction. As we have seen previously in this book, there is no empirical study of the short- and long-term effects of exercise for eating disorders sufferers, and due to methodological and ethical difficulties, it is hardly conceivable that a meaningful study of this kind could be provided. Partly, as we said, there are methodological issues relating to the difficulty in recruiting large numbers of eating-disordered exercisers willing to be monitored over a sufficiently long period of time. More serious ethical concerns relate to the exposure of these clients to possible preventable risks.

I shall begin this analysis by summarizing a short list of the main reasons for and against allowing eating disorders sufferers to exercise.

The main reasons for allowing people with eating disorders to exercise, in brief, are:

- The eating disorders sufferer is typically intelligent and functional. Paternalism is more difficult to justify when dealing with intelligent people, capable of understanding the consequences of their actions, and who only harm themselves.[6]
- The majority of fitness and sports instructors allow people with known eating disorders to exercise (except in the most severe cases).
- It is very difficult to ask for discontinuation of physical activities.
- Eating disorders are a medical problem, and requesting suspension of activities is not up to fitness and sports instructors.
- Suspending the activities of an eating disorders sufferer might be a form of unjust discrimination.
- Participation in physical activity might be beneficial to the sufferer.

On the other hand, however, physical activity can be harmful to the exerciser, and it is not part of the fitness professional's work to comply with the wish to use exercise to the detriment of one's health. The idea that someone could suffer serious injury, or even die, and that by allowing them to take the risk, we have somehow contributed to bringing this about, is a prospect that reasonably argues against allowing the exerciser to do more than they should.

There are thus strong reasons for and against accepting eating-disordered sufferers in supervised sports and exercise. I will now examine the strengths and weaknesses of these various arguments. From the examination, I will conclude that, in cases of serious and imminent harm, there is both a moral and a legal obligation to request suspension of physical activities. In less extreme cases, there are good moral reasons to accept the person with eating disorders. However, these good moral reasons do not generate a moral obligation for the fitness professional. Accepting clients at risk who could somehow benefit from exercise is a *supererogatory* act, something over and above their professional and moral duties. Ideally, qualified personnel should monitor the exertion of these clients.

In order to understand the ethical complexities of this case, it is important to clarify some ethical concepts.

Paternalism

Interfering with the freedom of others for their own good and without their consent, in ethics, is called 'paternalism' (see Chapter 7). Some believe that it is ethical to prevent others from making self-harming choices. For example, the philosopher Tom Beauchamp argues that paternalism is justified if it is likely that individuals will do serious harm to themselves or if they will deny themselves important benefits.[7] Thus, in these cases, a 'paternalist' might argue that, because the eating-disordered exerciser exposes her/himself to serious harm, there is sufficient justification for interfering with her/his freedom, and requesting a suspension of activities.

This view is called Strong Paternalism,[8] because it does not consider whether people are autonomous in making self-harming choices. The main pitfall of Strong Paternalism is that, if applied consistently, it would deny many people a great deal of personal freedom. People are free to choose activities and lifestyles according to their own preferences and needs, even if harm (or death) will result. People have the right to refuse medical treatment, even if they will die as a consequence of their choice.[9] People can undertake risky sports and jobs, live in unsafe areas, and so on. Strong Paternalism, if consistently applied, would severely curtail people's freedom in a way that is difficult to defend (especially in a liberal state or society).

Some restrictions on freedom for our own good are generally accepted: the laws on seat-belts, on use of helmets on motorbikes, on compulsory education, can be examples of these. In liberal states, however, intrusions on people's private choices are, or should be, kept to a minimum, and people have great discretion over matters that relate to their own body: diet and exercise certainly are included among these.

Healthy diets cannot be legitimately imposed, for example. Limitation of freedom to harm ourselves can be a violation of bodily integrity and individual autonomy, which are fundamental human rights secured by most conventions and declarations of human rights, and by the law of most liberal states.[10]

One might conceivably argue that the preservation of life has greater moral importance than respect for autonomy, but this presupposes some fundamental objections to the ethical principles and human rights that are widely accepted in liberal states. In other words, one might argue that liberal states are wrong, either because people are not really autonomous agents, or because there are stronger moral imperatives that bind us to protect and preserve people's lives, regardless of their own wishes and views. But clearly, an argument of this sort would have to be persuasive enough to sway a general assumption that people are sufficiently autonomous to govern themselves, and a widespread wish that people have to direct their own life. If one argues that it is right to force someone to eat in order to keep her alive, then one would also have to argue that it would be right to force someone to quit smoking, in order to prevent the onset of lung cancer, or to force life-prolonging medical treatment on someone who competently refuses it.

In principle, restriction of freedom to prevent self-harm collides with one of the pillars of liberalism, as stated in the nineteenth century by one of the fathers of liberal thought, John Stuart Mill:

> The only purpose for which power can be rightfully exercised over any member of a civilised community, against his will, is to prevent harm to others. His own good, either physical or moral, is not a sufficient warrant. He cannot rightfully be compelled to do or forbear because it will be better for him to do so, because it will make him happier, because, in the opinion of others, to do so would be wise, or even right. These are good reasons for remonstrating with him, or persuading him, or entreating him, but not for compelling him, visiting him with any evil in case he do otherwise.[11]

The strongest ground to justify paternalism is lack of autonomy on the part of the self-harming person. Mill argued that when someone is harming themselves while acting non-autonomously restrictions of freedom are justified:[12]

> If either a public officer or any one else saw a person attempting to cross a bridge which had been ascertained to be unsafe, and there were no time to warn him of his danger, they might seize him and turn him back, without any real infringement of his liberty.[13]

Here there is someone whose life is at risk, and who might be unaware of the risk they are running. Mill suggests that it is ethical to seize him and prevent him from walking over the unsafe bridge, in order to protect him. The justification for paternalism is not that he is harming himself, but that he is doing so while not knowing what he is doing. He lacks autonomy.

The law (in the UK, but similar provisions are found in most liberal states) provides that when a person's capacity to make a specific decision is jeopardized, either temporarily or permanently, s/he can (or should) be prevented from self-harm resulting from their lack of capacity. Interference with freedom, motivated by the fact that the person is lacking capacity or autonomy, is called Weak Paternalism.[14] The notions of autonomy and capacity/competence are at the center of complex debates both in philosophy and in law. This is not the place to analyze in depth the complexities surrounding these concepts. Suffice to say that in most liberal states, what is called Weak Paternalism is generally regarded as ethically legitimate, whereas Strong Paternalism is much more controversial, as it appears to violate the fundamental right to the exercise of personal autonomy.

Strictly speaking, restrictions of freedom aimed at the protection of third parties are not paternalistic. For example, excluding a person from a sports center because they have a severe infectious disease, or because they are abusive to others, is not an example of paternalism. These can be justified in terms of duty to protect public health or right to self-defense.

Key points

- *Paternalism*: interference with someone's freedom, in order to protect his or her welfare.
- *Strong Paternalism*: freedom is restricted, even if the individual is capable of making a decision for him/herself .
- *Weak Paternalism*: restrictions take place only when the individual is incapable of making a decision for him/herself.
- *Restrictions to freedom aimed at the protection of others* are not paternalistic.

Respect for autonomy: Eating disorders and exercise

John Stuart Mill declared, with words that have become the anthem of liberal thought, that "Over himself, over his own body and mind, the individual is sovereign".[15] The value of autonomy is universally recognized, secured by the constitutions of many states and subscribed to by virtually all declarations and conventions on human rights, such as the Universal Declaration of Human Rights (1948); the Convention for the Protection of Human Rights and Fundamental Freedoms (1950); and the Convention on Human Rights and Biomedicine (1997). The value of autonomy is also often stressed in both moral/political philosophy and medical ethics and is protected by law in several contexts as well as in health care.

Isaiah Berlin expresses this value in a famous passage:

> I wish my life and decisions to depend on myself, not on external forces of whatever kind. I wish to be the instrument of my own, not of other men's acts or will. I wish to be a subject, not an object; to be moved by reasons, by conscious purposes, which are my own, not by causes which affect me, as it were, from outside. I wish to be [. . .] a doer – deciding, not being decided for, self-directed and not acted upon by external nature or by other men as if I were a thing, or an animal, or a slave incapable of playing a human role, that is, of conceiving goals and policies of my own and realising them. [. . .] I wish, above all, to be conscious of myself as a thinking, willing, active being, bearing responsibility for my choices and able to explain them by references to my own ideas and purposes.[16]

People have freedom of religion, speech, and choice in most matters of private life. The protection of the welfare of the individual is in general not considered a sufficient reason to exert power over others and to force them to behave differently. Liberal societies generally support and defend people's autonomy in most areas of their life, so far as they do not directly threaten other people with their behavior, or limit the equal liberty of others.

From this point of view, it may be argued that everyone has a right to non-interference over his or her private choices relating to where, how, and how long for s/he wants to exercise and how much or little s/he wishes to eat.

It seems to follow that fitness and sports instructors might legitimately inform the participant of the risks they would run, if they exercised while suffering an eating disorder, but if the participant is capable of understanding the risks, it is up them to decide whether or not to exercise, and how and for how long. This may be regarded as an application of the doctrine of respect for individual autonomy in the area of sports and physical activity.

The case of Paula, case history 15 in Chapter 7, is an example of respect for autonomy in the fitness context. Paula argues that Lory, an eating disorders sufferer, has to *make up her mind* about exercise. She is informed, Paula can try to persuade her, but she cannot force Lory to do otherwise.

Key points

- Respect for autonomy is one of the fundamental pillars of liberal societies, and it refers to people's moral and legal right to direct their life without undue interference from others or the State.

One objection

As we saw earlier in this chapter, one ground for legitimate paternalism is lack of autonomy, or incapacity/incompetence.[17]

It could be argued that the case of eating disorders represents precisely an instance in which 'respect for autonomy' is inapplicable. Exercise, so the argument may go, is 'a symptom' of eating disorders, and as such cannot be an autonomous choice, one that others have a moral duty to respect. The eating disorders sufferer exercises not because s/he genuinely wants to do so, but as a result of an underlying mental illness that compels him or her to exercise. Respect for this behavior is not respect for people's autonomy: it is rather indifference to the fact that the person is ill.

A response: The objection is circular

The argument that anorexia nervosa causes people to diet, to fear fat, to over-exercise and so on, is a circular argument. This argument has been used in UK courts. For example, in *Re W*, the case of a legal minor refusing life-saving artificial nutrition, it was argued that anorexia nervosa "creates a compulsion to refuse treatment or only to accept treatment which is likely to be ineffective. This attitude is part and parcel of the disease and the more advanced the disease, the more compelling it may become."[18] In *Re C*,[19] another case involving a minor, the judge held that C was not competent to make a decision about the treatment for her mental disorder, *due to the effect of her illness*. These people were treated against their will not because they were minors, but because it was held that mental illness jeopardizes their decision-making capacity.

This view is that the person lacks autonomy *because of the mental illness*. Or, rather, mental illness *jeopardizes autonomy*. This argument is powerful and has held a consensus in ethics and law. However, it is a circular, and therefore flawed and untenable, argument.[20]

This argument is based on a mistaken assumption: the existence of a mental illness that produces effects, such as fear of fat, or desire for thinness. The idea here is that mental illness is a hidden something, somewhere inside the person, which produces emotions (fear, for example) and behaviors (diet or excessive exercise, for example). We have seen earlier in the book how many researchers have studied the physiology and genetics of eating disorders, and there is no conclusive evidence that an illness, or a dysfunction, of physiological imbalance or defect, can be responsible for the anomalies in behavior that we find in eating disorders. Indeed, many psychiatric categories, including eating disorders, are only names given to certain phenomena, and are not real things inside the person. The mental conversion through which an abstract concept[21] is thought of as a thing is called *reification*. Reification is *a logical mistake*, as it assumes that something exists when it does not.

Comparing anorexia nervosa with a medical condition can help to clarify this. Consider Alzheimer's disease. This is also regarded as a psychiatric illness. When memory loss appears, one could say, "memory loss is a symptom of Alzheimer's disease". In this case the claim makes sense. Alzheimer's disease is of course a *name* that summarizes a number of occurrences, but it is more than a summarizing name. It, in fact, also indicates the real presence of a degenerative process that affects the brain, and which causes a number of ill-effects, among which is memory loss. Memory loss is legitimately said to be a symptom of that process. When someone asks: "Why is this person forgetful?", a proper answer would be: "Because he has Alzheimer's."

Weight loss is not a symptom of anorexia in the same way as memory loss is a symptom of Alzheimer's disease, because, in anorexia, there is no tangible condition (either dysfunction, or degeneration, or physical anomaly) within the individual, which produces the effect of dieting. Weight loss is a result of dieting, not of anorexia. There is no underlying condition that produces dieting. The abnormal pattern of behaviors and experiences is all there is to eating disorders. A person receives a diagnosis of anorexia when (and because) she exhibits certain experiences and behaviors: excessive diet, overwhelming preoccupation with weight and so on (see Chapter 1).

The person is said to have anorexia *because* s/he exhibits those patterns of behavior; to say that she exhibits those patterns of behavior because s/he is anorexic is circular. This circularity is in philosophy referred to as *tautological*. A tautology is an empty statement. To say that a person fears fat, diets and exercises excessively and so on *because she has anorexia* is tantamount to saying that a person fears fat, diets and exercises excessively and so on, *because she fears fat, diets and exercises excessively and so on*, as having anorexia means no more than *fearing fat, dieting and exercising to excess and so on* . . .

Anorexia, and most other psychiatric categories, are descriptive items. Of course Alzheimer's is also a name, and most other illnesses are given names. Sometimes they are given the name of those who discovered the causes of the disease. But Alzheimer's, like many other names of pathologies, are not just names: they refer to the biological variations that both cause and explain people's abnormal experience and behavior (why they have pain, for example; why they can no longer see properly, and so on). The names that indicate eating disorders instead only refer to the disorders in eating. These names (anorexia, bulimia, binge eating, eating disorders) summarize, in single words, a whole pattern of behavior and experiences, but say nothing about their causes nor how they might be explained. This fact is often overlooked, and it is taken for granted that if someone suffers from eating disorders, it is for that reason that they diet and put their health at risk for the sake of thinness. This is the flaw in the argument. The description is taken for an explanation. The name is thus taken as 'the underlying condition', when there is no proof that any underlying condition actually exists which causes the stated effects.

This leads to a circularity, which is linked to the process of reification explained above, which is commonly found in discourses around mental illness. For example, it is said that people fear open space *because* they are agoraphobic. However, being agoraphobic means *fearing open space*. Thus, that statement amounts to saying that people fear open spaces because they fear open spaces – given that being agoraphobic means fearing open spaces. Or it is said that one cannot control gambling because she suffers from a gambling pathology. If suffering from a gambling pathology means 'being unable to control gambling', then saying: 'she cannot control gambling because she suffers from pathological gambling' is like saying that a person cannot control gambling because she cannot control gambling.

To say that a person diets because she has anorexia is as nonsensical as to say that a person gambles because she has pathological gambling, or that a person has panic attacks because she is phobic.

Psychiatric categories have an important *predictive value*. This means that, when someone introduces a person to me, saying that she is anorexic, I expect to see a thin lady, who might refuse to eat my cheesecake. The psychiatric diagnosis is also important in other ways:

- It helps to 'set things into a context' (one senior colleague, Harry Lesser, once told me that, when his son was young, the TV reported the news of a TV presenter committing suicide. When the child asked him: "Why did he kill himself?", Lesser answered: "Because he suffered depression." In doing so, Lesser aimed at setting the presenter's action into a context: it was a short way of saying that committing suicide is an extraordinary event, that happens only in certain rare circumstances. In this sense, a psychiatric diagnosis can help to 'set things into a context');[22]
- It makes people's self-harmful choices and actions more tolerable both to others and themselves;[23]
- It helps everyone avoid judgmental attitudes towards the sufferer (in spite of the stigma associated with mental illness);
- It releases the sufferer from feelings of guilt.[24]

However, the diagnosis has no explicative value, and claims that people diet to excess and over-exercise *because they have eating disorders* are circular and meaningless.

Key points

- Claims that people adopt abnormal eating and exercise because they suffer from eating disorders are nonsensical.
- Claims that anorexics are not autonomous because they have a mental illness are nonsensical.
- Claims that people are unable to control their diet and exercise because they suffer from eating disorders are nonsensical.
- The ethically related claim that it is right to restrict their freedom because they cannot be autonomous because their behavior is driven by mental illnesses is unjustified.

Does this mean that eating-disordered exercisers are autonomous?

It might of course be true that people with eating disorders lack autonomy in some important way, even if it is mistaken to claim that *eating disorders* jeopardize their autonomy. Yet it is important to refute claims that *eating-disordered people lack autonomy because of their mental illness* or similar. These claims have in fact been used in the attempt to defend severe restrictions of freedom in the English courts.[25] If my arguments are persuasive, then any attempt to morally or legally justify coercion on that basis fails.

If one believes that the eating disorders sufferer lacks autonomy, then they need to demonstrate it. Even people with profound suffering, and with a full diagnosis of eating disorders do not 'lack autonomy' only by virtue of their diagnosis. Appealing to a psychiatric diagnosis is insufficient.[26] Many have indeed embarked on various complex attempts in the 'hope' that they will demonstrate that people with eating disorders lack autonomy, and thus coercion can be ethically justified. But their attempts have not been convincing.

For example, many (see Chapter 1 for a full account) have argued that people with eating disorders have a distorted body image. If someone sees themself as three times as big as they really are, you might argue that they cannot diet autonomously. But no study has provided sufficiently persuasive evidence of this widespread claim that eating disorders sufferers have distortions in the body image. Others have claimed that prolonged dieting is detrimental to cognitive abilities, and those with impaired cognition cannot make competent decisions. But again, nobody has been able to demonstrate this.[27] As we saw at this beginning of this book, many have also tried to prove that eating disorders are addictions, or caused by genetic disorders. But no study has demonstrated yet that the eating disorders sufferer has impaired reason.

If there is no proof that all eating-disordered exercisers lack autonomy in some relevant way, then restrictions of freedom cannot be justified on the basis that the person's autonomy is jeopardized. Weak paternalism, in other words, does not seem to be a justified option with eating-disordered exercisers.

One alternative argument is that restrictions of freedom are justified under the principle of beneficence. The argument might go like this: the eating-disordered exerciser might be autonomous, and can make competent choices about diet and exercise. Yet, there is a *prima facie* moral obligation to protect other people's welfare, or to minimize harm, and therefore requesting the suspension of physical activity is, at least, a morally legitimate thing to do (if not even a morally dutiful thing to do), in order to benefit the client and avoid preventable harm.

As we are now going to see, however, from the point of view of beneficence, things are not so straightforward.

Beneficence in fitness practice

The principle of beneficence suggests that people have a moral duty to do what is good (*bene*, in Latin) for others. The extent and limits of this duty can be debated. For example, I cannot be morally obliged to cause great pains and discomfort to myself in order to give a minimal benefit to another. But it is debatable whether I ought to endure significant pains and discomfort in order to give a maximal benefit to another, for example, to save his life. And perhaps I ought to endure a minimal discomfort to myself to give a maximal benefit to another.

An analysis of the theoretical issues relating to the scope of beneficence is not necessary for our present purposes. Earlier in this chapter, we saw that restrictions of freedom of action, when the action is autonomously or competently made, for the good of the person concerned, are difficult to justify. But let us assume for a moment that it would be ethical to do what is good for others regardless of what they think, or against their wishes. Still, it would be necessary to understand what is 'good' for others. In the fitness context, it is clear that fitness professionals, as health professionals, have an imperative to promote people's health and to do no harm (and this includes *prevention* of harm – see Chapter 7). The main function of fitness professionals is to enhance people's fitness, which is one of the fundamental components of health and welfare. Doing good in this sense is inherent to fitness practice. As the following sections show, contrary to what one might initially think, beneficence can actually require that we allow eating-disordered exercisers to take physical activity. Beneficence might require that, instead of acting paternalistically against the person's wishes, we comply with them.

Beneficence as minimization of harm

Contrary to what might appear, beneficence might suggest that we accept the eating-disordered exerciser under our supervision, in spite of the risk to his/her health. Let me clarify this with a case from my own experience.

One day I had a clearly severe anorexic in my class. I was profoundly bewildered and asked myself what I should do. My first thought was: "She can't exercise, she'll collapse." My second thought was: "If I ask her to leave, she might go running in the park, and if she collapses, nobody might be there to rescue her. If she collapses here, we can give professional assistance and call an ambulance straightaway." This seemed to me at the time a good enough reason to allow her to take my class. In Chapter 7, we have seen that some other fitness professionals have used a similar reasoning. For example, in case history 8, Alex said: "If you refuse an application, the client will enroll at another gym." The idea is: I know I am going to care for her, and I don't know how others are going to care for her, so she'd be better off with me.

This stream of thinking, in philosophy, is called *minimization of harm*. This is one of the most widely accepted principles in ethics. The philosopher Karl Popper wrote:

> I believe that there is, from the ethical point of view, no symmetry between suffering and happiness, or between pain and pleasure. Both the greatest happiness principle of the Utilitarians and Kant's principle, 'Promote other people's happiness . . .', seem to me (at least in their formulations) fundamentally wrong in this point, which is, however, not one for rational

argument. [. . .] In my opinion [. . .] human suffering makes a direct moral appeal for help, while there is no similar call to increase the happiness of a man who is doing well anyway.

'Minimize suffering'. Such a simple formula can, I believe, be made one of the fundamental principles (admittedly not the only one) of public policy. (The principle 'Maximize happiness', in contrast, seems to be apt to produce a benevolent dictatorship.) We should realize that from the moral point of view suffering and happiness must not be treated as symmetrical; that is to say, the promotion of happiness is in any case much less urgent than the rendering of help to those who suffer, and the attempt to prevent suffering. (The latter task has little to do with 'matters of taste', the former much.)[28]

Smart and Williams explain that "by 'suffering' we must understand misery involving actual pain, not just unhappiness".[29] It has also been recognized that "in most cases we can do most for our fellow men by trying to remove their miseries".[30] Virtually all moral codes include some principle of non-maleficence, that is a principle of not harming or choosing the course of action that is least harmful.[31]

The eating disorders sufferer who is refused access to the class will probably not renounce exercise: s/he may go running, or cycling, or swimming, instead.[32] Exercise on an individual basis is probably overall more risky for the sufferer than supervised exercise. Supervised exercise is in fact generally structured so that the participants can exert themselves within safe parameters.[33] A number of methods (observation, music choice/pitch, talk test, perceived rate of exertion, heart rate monitoring) may be used to keep under surveillance the participant's level of workout.[34] Therefore, supervised exercise can be considered safer than other types of exercise, in which quality and appropriateness of movement, and heart rate, are not checked, or not checked by qualified staff. Moreover, in the case of an accident, the injured person would probably receive better and quicker help in a sports center. The principle of minimization of harm allows a comparison of harm: harm that might result from unsupervised exercise, harm that the exerciser might suffer when s/he is asked to suspend physical activities, versus the potential harm that s/he might suffer in a supervised context.

From the point of view of minimization of harm, professionals have good ethical reasons (or maybe even a moral obligation) to accept eating-disordered participants.

However, in spite of its intuitive force, minimization of harm has important weaknesses.

Weaknesses of minimization of harm

Let me explain why I think I was wrong when I allowed the severe anorexic into my class.

First, if my reasoning was valid, it would follow that doctors should, for example, prescribe heroin to drug addicts every time they threaten to harm themselves outside their clinic should they not receive the drug. It is not clear whether health professionals have a moral obligation to offer harmful medications in order to prevent possible greater harm elsewhere. Certainly, clinical prescriptions should not be dependent on similar threats. Likewise, it is not clear that fitness professionals should allow people to harm themselves in the fitness center in order to prevent them from suffering potentially greater harm elsewhere, especially when there are safer alternatives (for example, as mentioned in previous chapters, direct negotiation with the client, involvement of social services, liaison with health care professionals, helplines or discussions with senior or more experienced colleagues).

Second, there is an issue of responsibility, and one of probability, which the arguments for minimization of harm do not fully consider. I am surely responsible (morally and professionally) if I let the anorexic continue to exercise in my class. But I am not responsible in the same way if the anorexic goes running against my advice. There is a wider ethical issue of moral responsibility here: in what sense are we responsible for what we bring about with our decisions? We are arguably morally responsible for what we contribute to bringing about with our decisions: this is why we need to think about what we should do, and we need to do it with tact and in a reasonable and sensitive way. Of course, we should never do or say anything that the client might reasonably perceive as punitive or hurtful. Should a client walk away from our center feeling hurt and humiliated, of course the instructor would be to an important extent morally and professionally responsible for this. But the moral and professional responsibility that we have for someone who chooses to train outside the sports center, outside the trainer's supervision or against the trainer's advice cannot be the same as the moral and professional responsibility that the trainer has towards the supervisee. The minimization of harm argument overlooks this difference. There is also an issue of probabilities to address: the harm resulting from the sufferer running elsewhere is speculative, whereas the harm that s/he would suffer in the center under supervision is real and present. Preventing a perhaps higher degree of harm that is merely possible is not a good reason to bring about a foreseeable and likely accident.

Finally, if fitness professionals are to properly manage the risks (balancing the risks of the eating-disordered exerciser taking supervised or unsupervised physical activities), they should be properly qualified to do so, or otherwise work together with health care professionals. Proper minimization of harm can take place as a part of a wider program, but not on its own.

Eating disorders, exercise and minimization of long-term harm

Claire, in case history 11, argued that it was wrong to bar the anorexic sufferer. Exercise could not really harm her any further, whereas being barred could be traumatizing for her. After she was asked to leave the fitness facility, her health degenerated, and Claire argues that the exclusion from the gym might have caused this deterioration. Claire was thinking about the general and long-term harm that might result from being barred from the fitness center.

Indeed, limitations of freedom (to exercise, for example, but other limitations might be forceful feeding or involuntary hospitalization) might be particularly harmful to people with eating disorders. One of the psychological dynamics precipitating eating disorders is the sense of lack of independence and control. It is accepted that extreme diet is the expression of the sufferers' profound need to perceive themselves as capable of self-determination.[35] For this reason, paternalistic acts towards anorexics are generally seen as compromising long-term recovery.[36] It has been argued that, whereas physicians always confront potentially conflicting responsibilities (respecting people's wishes and promoting their welfare), in the case of eating disorders they have "to face an additional *therapeutic* dilemma: encouraging greater autonomy and freedom in the patient, while at the same time taking control over the patient's food intake and body weight".[37] Some famous clinicians and scholars of eating disorders talk of "self-determination theory",[38] according to which respect and promotion of patient autonomy is essential to recovery. They propose a therapeutic approach based on this theory: patients are allowed to decide on their nutrition and treatment, and negative responses are avoided when they choose not to eat. They conclude that when patients are given choice, the drop-outs from treatment are significantly smaller.

Paternalistic acts might thus have counterproductive effects.[39] They might erode further the already fragile self-esteem of the sufferer. They might also erode the trust in the professional, which is necessary for improvement. Paternalistic interventions are likely to render the sufferer even more aggressive and determined in her behavior, more frustrated, angry and lonely than before. Acts of paternalism in the management of eating disorders are therefore a particularly delicate choice, probably more delicate than in the management of other conditions. The implications of this for the fitness profession are clear: the professional who is *primarily* concerned with minimization of long-term harm has good reasons for allowing eating-disordered exercisers to take supervised exercise. The argument for minimization of long-term harm also has weaknesses, however, that will be explored later. Exercise, moreover, can promote long-term good, as I am now going to argue. I will now explain further strengths and weaknesses of the principle of beneficence. Later, I will argue that contractualism might provide a better framework for decision-making involving eating-disordered exercisers.

Beneficence as maximization of long-term good

In case history 7 (Chapter 8), Y. narrates the story of one of her students, a young gymnast, who has anorexia nervosa. She relates her condition to her family history.

> The one who is clearly anorexic, for example, has a terrible family situation, we would never abandon her. She has lost her mother when she was only three, her father is with another woman and she doesn't feel very much loved and accepted by this woman. She is very attached to us, and if we reject her, she will hurt a lot. We have had her for years now, since she was maybe five

or six, and I am also attached to her. It never really crossed my mind seriously to tell her father to remove her, I just wouldn't do it.

She points out that, as a gymnastic instructor, she feels she can do a lot of good and, potentially, a lot of harm. Should she bar her young client from the gym, she feels the athlete would suffer great long-term harm, whereas she feels confident that accepting her and working with her will eventually help her to gain a good relationship with food and with her body. "Gymnastics can help you to love yourself. You can turn exercise into a way of caring for yourself and loving yourself. It's the coach who has to be good enough for that." Exercise can indeed be used as a part of rehabilitation, and some therapists use physical activity in this way.[40] Exercise can help to increase self-esteem, to reduce anxiety, and avoid the stress of having to stop physical activity due to the disorder. There is also evidence, as reported in Chapter 6, that athletes with subclinical eating disorders who are forced to stop activity tend to reduce food intake to compensate the decrease in daily calories burned out.

Fitness professionals who are mainly concerned with long-term health gains might feel that they have a moral obligation to accept eating-disordered clients in their classes, that they must accept them. Despite the fact that instructors may have apparently cogent reasons to accept eating-disordered participants, I will now argue that they are not morally obliged to do so.

Key points: Ethical reasons to let eating-disordered participants take part in supervised sports and exercise

- **Autonomy:** Respect for individual choices (assuming those choices are genuine).
- **Beneficence:**
  - **minimization of short-term harm:** protection from more serious consequences – exercise in solitude and without specialized supervision;
  - **minimization of long-term harm:** prevention of sense of lack of control and autonomy;
  - **maximization of long-term good:** promotion of sense of self-control and responsibility and maximization of long-term health.

The implicit contract between fitness professional and client

In short, there are five different, though not mutually exclusive, ways of thinking about the moral dilemmas raised by the eating-disordered exerciser.

- First, there are those who think that their primary moral obligation is to respect people's wishes (*respect for autonomy*);
- Second, there are those who believe that their primary moral obligation is to prevent people's short-term harm (*minimization of short-term harm*);
- Third, there are those who believe they should prevent long-term harm (*minimization of long-term harm*);
- Fourth, there are those who believe that they should promote long-term good (*maximization of long-term good*);
- Fifth, there are those who believe that accepting an eating-disordered exerciser is not a part of the contract between clients and professionals (*contractualism*).

It is to this fifth way of thinking that I now turn.

Case history 10 is an example of *contractualism*. Ruth argued that it is right to request suspension of activities in cases of severe eating disorders because fitness instructors are not qualified to deal with severely debilitated exercisers, and cannot be asked to cooperate with exercisers who want to harm themselves. This is a form of *contractualism*, which seeks the answers to moral dilemmas in the understanding of the implicit moral contract between professional and exerciser.

As discussed in Chapter 7, contractualism is a term of political philosophy. Its origins can be traced to the works of Jean-Jacques Rousseau, and in particular in *The Social Contract*. Other philosophers, like Immanuel Kant, in the 1700s, and more recently, John Rawls and Tim Scanlon, have articulated the theory of contractualism. Their theories are mainly meant to explain the fabric of civil society, but they can also be understood as theories of morality. For example, for Rousseau the moral action is the one which, within the social sphere, is acceptable to others *as equal*. For Kant, instead, it is the one that reflects the rational nature of humans and that obeys universal moral principles, regardless of the particular features and sensitivities of each individual (including the agent). Hobbes produced a different idea that is called contractarianism, in which individuals act socially in their mutual self-interests. Rawls tried to find the basis of morality in a contract similar to that proposed by Kant, in which individuals, ignorant of the place they have in society (under a veil of ignorance), choose on the basis of what an ideal agent would choose under the circumstances. And Scanlon argued that the basis for moral action resides in the rules that each rational agent has reasons not to reject. One idea that is common to various theories of contractualism and contractarianism is that morality resides in the original agreement made by those in the moral domain.

I argue this is the most useful stream of thought, among all those analyzed so far, in the management of eating disorders in sports and exercise. In explaining why this is the strongest argument, I will also clarify why *minimization of long-term harm* and *maximization of long-term good* are not fully satisfactory arguments. I have already discussed above the flaws of *minimization of short-term harm*.

Accepting a participant with eating disorders is tantamount to accepting *a different job*. Accepting any participant is, in essence, a private transaction, and this is not unique to eating-disordered exercisers, as far as contractualism is concerned. The eating-disordered participant asks the instructor to enter an *atypical* contract with them. Accepting the supervision or training of an exerciser with eating disorders means entering a special type of contractual relationship with the participant. Nobody can be obliged, nor yet morally obliged, to accept a different job, especially one in contrast to a proper understanding of the function one has as a fitness professional. Duty of care, discussed in the previous chapter, requires that foreseeable harm to the health and safety of the client be avoided. In cases of severely disordered patterns, our duty of care requires that we address the problem and, if negotiation is impossible, that physical activities are suspended.

In order to understand why we are not obliged to allow an eating disorders sufferer to take supervised exercise, despite the arguments from beneficence, we need to gain a proper understanding of what is involved in fitness practice and how we should understand our role. Fitness practice has three main characteristics:

1   Fitness is a customer service;
2   Fitness activities are tailored for reasonably healthy adults (with exceptions discussed later);
3   Fitness activities involve and are based upon risk assessment.

Fitness is a customer service

Fitness is a 'customer service'.[41] Customer service is defined as "the provision of service to customers before, during and after a purchase".[42] Customer service is offered in various formats (by representatives or even by automated services – e.g. the internet) by companies to show that they provide more than mere 'products'. Customer service is thus the corollary of attention that is dedicated to customers, or clients, before and/or after purchase, to increase their satisfaction and improve the company's credibility on the market.

The major difference between many other customer services and fitness services is that whereas the company normally knows what it is selling, fitness professionals do not know what they are selling. Or better – they might have a clear idea of what they are selling, but might not know what the customer is hoping to buy! People participate in fitness activities for a multitude of reasons, and therefore professionals end up selling an indefinite number of things – for some, sports and fitness activities are mainly social activities, a way to meet up with friends or getting to know new people; for others, they are ways of improving general health; others go to the gym to lose fat; others to develop muscles; others to prevent deterioration of existing medical conditions; others just go 'for fun' and so on. Every individual comes to the sports and fitness center with their own expectations and hopes. Because every client 'buys' something different, this customer service could be described in many different ways. What matters here is not attempting to explore these various ways and finding the most suitable description of fitness services, but to understand the core elements of what is bought and sold. Central to any possible characterization of fitness services, there surely must be that it provides *benefits for health* (a full account of the benefits of exercise for health can be found in Chapter 6). This is part of the codes of ethics: this is what professionals should do, provide fitness and health.

Exercise is not beneficial to the person with eating disorders as it is to the averagely healthy individual, and may even threaten her/his life. Although a perhaps bigger potential threat exists for the person *outside* the class, exercise in the class also involves *direct risk* for her/his health and life. Fitness professionals, thus, by accepting the eating-disordered exerciser renounce the provision that they are supposed to provide – a service aimed at improving health and welfare[43] – and agree to work for goals that are alien to any appropriate understanding of their job.

It is true that professionals might minimize short- and long-term harm, and maximize long-term health by accepting the eating-disordered exerciser; but by doing so they would take on responsibilities that are not part of their job as fitness and sports professionals. This would not necessarily and always be unethical; it would just go beyond the call of duty.

The Fitness Standard Council Code of Ethics, at point 15, and other US codes of ethics also stress (see Chapter 8) that fitness professionals should not attempt to undertake tasks that are beyond their qualifications and expertise.

Moreover, long-term harm resulting from restriction of freedom might be to some extent speculative, because we cannot always predict with certainty what will happen to a person outside our center and in the future, and how frustration of freedom, at this stage, will affect his or her overall welfare. Instead, the harm to which we agree to expose the eating-disordered exerciser in our facilities is real and present. This harm is something that fitness professionals should refuse to sell.

It should also be noted that fitness professionals who are concerned with what may happen to the participant outside the class may have other options – advice help lines, suggesting a specialized clinical psychologist, discussing the risks to health with the participant, contacting local social services – that do not involve a direct risk. For all these reasons, and others that I shall discuss now, *contractualism* gives a better understanding of the fitness professional's ethical responsibilities.

One objection: How about respect for autonomy?

One could raise an objection: does this mean that fitness professionals do not have to respect people's autonomy? I have suggested that there are similarities between the ethical dilemmas encountered by some health care professionals and those encountered by fitness professionals. Healthcare professionals are (or should be) also primarily concerned with people's welfare, and yet they must respect people's choices, even where they are detrimental to their patients' health. Why am I applying different ethical standards to fitness professionals? If health care professionals must respect people's choices, surely fitness professionals must too.

This objection, however, fails to capture the difference between situations in which clients request a service and situations in which clients refuse a service. Doctors have a stringent ethico-legal duty to respect people's autonomy *when they refuse medical treatment*. They do not have a similar stringent duty to respect people's *requests* for treatment. Likewise, fitness professionals cannot force a person to undertake a fitness program if s/he refuses to take part in physical activities. They can promote exercise and fitness, can advertise it, can try to persuade people to participate, but cannot, with force or threat, compel anyone to take exercise. Fitness professionals have discretion over who may become their client. There are limits to this discretion. The entitlement to refuse a client is not absolute, as the doctor's entitlement to refuse to treat is not absolute, and has to be justified on moral or clinical grounds. A client being of a different ethnic origin, or of a different religion, or having other arbitrary features that are not relevant to the fitness program are not acceptable reasons. Refusing to teach someone because they are black, or Muslim, can be examples of unjust discrimination. This could also expose the instructor to legal challenge. Refusing a client because s/he is at high risk due to his or her ill health is not unjust discrimination, because duty of care requires that the health and safety of the client are protected (Chapter 8).

Harm for sale

As I have also argued at greater length elsewhere,[44] health professionals are responsible for their omissions, not only for their actions. Omissions are not morally neutral islands.[45] This applies to sports and fitness contexts too. Where there is an imminent risk to the health and safety of the participant, the instructor should use her judgment and competencies to minimize that risk. So, for example, if there is strong evidence that the client will harm herself if she is asked to stop exercising, there is a forceful ethical reason not to ask her to do so – there is instead a strong ethical reason to do other things that might be appropriate in the given circumstances, such as not leaving the person alone, attempting to talk about the distress, asking the person if it is all right to contact a friend or a relative, and so on. Is this coherent with the idea that we cannot be morally obliged to accept tasks that go beyond the proper understanding of our function, as fitness professionals? This raises a number of other ethical questions: how responsible are we for what people choose to do? To what extent can we be held morally responsible for what happens to people as a result of what *we fail to do*? This problem in ethics is known as the 'acts and omission' puzzle. Are we responsible for our omissions? And if so, to what extent?

There is a further complication: what is 'an omission'? If I refuse to train my anorexic client, in a sense I *omit* to train her. But if I allow my anorexic client in my class, in a sense I *omit to* address her problem and give due importance to her physical condition. It is not entirely clear what should count as an omission as opposed to an action. This has been pointed out in other contexts, and has puzzled both philosophers and the courts, in some cases.[46]

This is my proposed resolution to these puzzles. My implicit contract with the client binds me, morally and legally, to offer *improvements* to her or his health, and not to harm the client. Sales of harm are not part of my implicit contract with the client, as it is correct to understand it. Where it is highly likely that the health and safety of the client is at imminent risk, I should take the course of action that prevents the occurrence of that risk materializing itself. In principle, we should opt for the course of action that avoids highly-likely harm rather than the merely possible. This moral duty falls on us not just particularly as fitness professionals, but also generally as decent people.

Fitness professionals move on the sliding scale of moral responsibility, and acting responsibly involves making an exercise of balance – this is especially true because they are not only *fitness leaders*, but also human beings, who can enter into human and complex relationships with their trainees. In principle it has to remain true that fitness professionals should sell health and fitness, and not harm. But at times we can be morally required to act beyond the call of duty *as professionals*, and to do things required by the moral 'rule of rescue': I do something I would normally not do, and something that I am not called to do *professionally*, but I do it because this saves a person from serious and imminent risk.

Even in these cases, the problem of the vulnerable exerciser should be addressed and there are various ways in which fitness professionals can do this: they can and should talk to the sufferer openly about the condition, advice the suffer to contact local social services, or their GP,[47] give information about eating disorders or helplines. These are all examples of how the problem can be tackled without infringement of the client's autonomy or breach of confidentiality. Ignoring the participant's ill state and letting her or him exercise as they wish is a neglectful omission, which is not acceptable within the profession.

Fitness activities are tailored for averagely healthy people

The second core element of fitness activities supplied in most sports centers is that they are tailored for averagely healthy people. Specific activities and classes are designed for particular groups (people with disabilities, elderly people, children, pregnant women, etc.), for which specific qualifications are required.

Of course, there will always be a degree of non-preventable risk in exercise. The instructor is probably unable to identify all eating-disordered exercisers, and people with other non-manifest or unknown medical conditions. However, there is in some cases sufficient evidence that an individual cannot exercise safely, evidence that may be obtained by use of basic observation skills. We have also discussed in previous chapters the signs and symptoms that can help professionals to identify participants at risk of having eating disorders. Interpretation of PAR-Qs and other written screenings, which should always be made available to any fitness instructor on request, can also assist in this process.

Eating disorders sufferers can certainly be considered less than averagely healthy (see Chapters 2 and 6), and it is therefore inappropriate to include them in activities designed and set up for averagely healthy clients.

Risk assessment

According to the National Occupational Standards framework for health and fitness, fitness instructors should be able to identify safety issues and perform adequate health and safety screenings. Determining the health and fitness status of an individual prior to the initiation of an exercise program is one of the most important roles of fitness instructors.[48] This is known as *health screening*, and is one of the forms of risk assessment that are a part of the instructor's job. Other types of risk assessment include:

- Assessment of the working space
- Assessment of emergency facilities
- Assessment of the equipment (a fuller account is found in Chapter 8).

Screening (see Chapter 8) helps professionals to set up adequate exercise programs, and also to recommend the client *to seek medical advice* if important risk factors are detected. As Mickley points out, "[i]f the instructor feels that the form reveals anything that is beyond the capacity of their training or experience, they should refer the case to a suitably qualified colleague or supervisor".[49] This of course does not mean that supervisors or better qualified colleagues will always have ready answers:[50] it means that the matter should not be ignored, and that it is important to cooperate in the framing of an action plan that is in line with ethical and legal responsibilities.

Understanding the three core features of fitness activities helps to appreciate why it cannot be mandatory, from an ethical point of view, to accept eating-disordered exercisers, and why in severe cases s/he should be asked to suspend physical activities.

When an emaciated person enters the gym or the studio, s/he implicitly asks the instructor to enter into a new and atypical (implied) contract. S/he asks the instructor to handle a situation for which the instructor is not qualified, and to renounce giving due importance to risk assessment. Professionals who have reason to believe that the person has an eating disorder and who decide to allow her/him in the class agree, by doing so, to comply with sets of goals and responsibilities that are alien to any proper understanding of their function as fitness professionals. This is not necessarily *unethical* (as it might be justified under the principle of beneficence). However, I have argued, it goes beyond the call of duty.

The relationship between fitness professionals and clients is a contractual relationship, and as such it must be voluntary. It must therefore remain possible, and it must be regarded as ethical, for the instructor to retain discretion over selection of clientele based on health and illness, and to exercise the legitimate freedom to request suspension of activities for those who are at risk.

If the fitness professional has reasons to believe that the client has a relevant condition that can be aggravated by exercise, s/he should take action. The NSCA Professional Standards and Guideline Task Force of Colorado suggests that if a fitness instructor "had prior knowledge of an existing danger or risk but took no corrective action to help prevent resulting injuries, this failure to act would most likely constitute an extreme form of conduct" (defined as gross negligence, willful and wanton conduct, or reckless conduct).[51] Risk assessment is one of the fitness professional's ethical responsibilities, and they should request discontinuation of activities based on their screening, if they believe that exercise can pose a significant threat to the client.

Professionals may be persuaded to accept eating-disordered participants or others at risk for the ethical reasons discussed above. By agreeing to care for what may happen to the person outside the class or in the future, fitness professionals partly renounce the responsibilities that are within the scope of their job, namely, caring for what can happen to the participant in the class and while exercising.

It should of course be recognized that fitness and sports professionals may feel unable to deal with the problem, and may believe that not taking any action is best. And of course, it feels horrible to ask someone with eating disorders to leave. However, fitness professionals are not there to serve their own interests, but those of their clients. Ignoring a serious problem because one 'feels bad' about it, is a form of self-indulgence that cannot form a sound basis for ethical action or good practice.

One could believe that the majority of fitness and sports instructors would allow people with known eating disorders to exercise (except in the most severe cases), and that it is appropriate to follow this general rule of thumb. There is no clear evidence of what the majority of instructors do in the given circumstances. Chapter 7 has shown that fitness professionals respond in different ways to these cases. But even where this was the case, this would be ethically irrelevant. What is relevant is what is reasonable to do in the given circumstances, and not what others would do. Others might be unreasonable.

One could believe that eating disorders are a medical problem, and requesting suspension of activities is not up to fitness and sports instructors. Nonetheless, fitness instructors should be able to make a judgment as to whether a person is fit enough to exercise, and request a medical certification where they are unsure.

Finally, as we saw above, one might be concerned that suspending the activities of an eating disorders sufferer might be a form of unjust discrimination. However, discrimination occurs when a client is treated differently from others on morally or legally irrelevant grounds. Health status is a relevant ground. There can be no case of discrimination when the instructor requests suspension of activities to a client who is at serious risk, due to his or her ill health.

Participation in physical activity might be beneficial to the sufferer. This can give moral reasons for accepting an eating-disordered exerciser in supervised physical activities, but this might go beyond the call of duty. Physical activity can in fact also be harmful to the client, and in case of serious and likely harm, duty of care requires suspension of physical activities. It is not part of the fitness professional's work to comply with the wish to use exercise to the detriment of one's health. The instructor is not morally obliged to undertake tasks that go beyond his/her competencies, and to comply with attempts to self-harm. Instructors have a good deal of discretion over who they reasonably think is able to participate in sports and fitness activities. If there is no reasonable way of knowing that the person has an eating disorder, the eventual harm is not imputable to the instructor. But if the instructor has reason to believe that the person is at risk, then duty of care requires that preventable harm to clients is avoided.

Key points

- When physical activity poses a serious and imminent risk to an individual's health, or in case of aggressive or indecent behavior, fitness professionals should request immediate suspension of physical activities.
- When unable to handle the problem, the professional ought to seek advice from better qualified colleagues, supervisors, studio coordinators or gym managers.
- Fitness professionals should not sell exercise that can harm the client.
- Communication with other services (social services, advice help lines, line managers, GPs) should be considered in cases of eating disorders or other potentially self-harming behavior.

Conclusions

Exercise is an important part of life for many eating disorders sufferers. This is why a large proportion of those with eating disorders exercise to excess. There are important moral reasons that can induce fitness professionals to accept eating-disordered exercisers. It is not unethical to do so, but it can go beyond the call of duty. Exercise can be turned positively into an instrument of self-worth, and in principle fitness professionals could play an important role in the transition from self-destruction to self-love. Despite this potential, current qualifications for fitness professionals do not include knowledge of the physiology and psychology of eating disorders, and thus the average fitness instructor is unprepared to monitor the activity of an eating-disordered exerciser. Thus the professional could expose the participant to serious risks, by accepting her or him among the supervisees.

There are several ways of thinking about the moral entitlements and obligations towards those at risk. I have argued that the strongest stream of thinking is a form of *contractualism*, and that the lines of actions subsumed under the principle of beneficence have flaws and go beyond the call of duty. Duty of care requires that people are not exposed to preventable harm. Fitness professionals must remain firm to their role as promoters and protectors of health and safety, and should refuse to incite harm, no matter how strongly the clients want it.

Contractualism does not bind the instructor to refuse his/her services in every case in which a client is less than averagely healthy. Instead, it binds him/her to make a considered judgment over whether s/he can satisfy the client's wishes without renouncing professional integrity.

10 Recommendations and conclusions

Introduction

This book has analyzed some of the major ethical and legal issues raised by the participation of people with eating disorders in sports and fitness. There are other vulnerable clients, with known and unknown medical conditions, and for whom the exercise they take might not be as beneficial as they might think, or as beneficial as it is for the generally healthy person. Although the focus of this book has been eating disorders, the conclusions gesture towards the extension of these arguments for sports people who exercise to the detriment of their health and those with other health problems who choose to exercise at risk to themselves.

One of the main functions of philosophical analysis is to enhance understanding of facts, values, and to unravel critical issues. It is often by shedding light on the terms or the elements of a dilemma that that dilemma can be resolved. I have thus analyzed eating disorders and attempted to identify their possible causes and triggers. I then discussed whether these forms of body abuse can be conceptualized as addictions, and what 'addiction' means. Finally, I have considered the role of the media in the spread of eating disorders, the historical, sociological and normative meanings behind the preference for thinness, and the significance of the exposure of the woman in the public eye.

What emerges from this exploration is that eating disorders are a complex condition, which might result from the interplay of many different factors (see in particular Chapters 1–5). Yet, people's deliberation and will are not ruled out in such conditions, contrary to what might seem to be the case and to what some might claim. There is no evidence that people with eating disorders are unable to understand the consequences of their behavior, and no evidence that they are unable to monitor their eating and exercise habits. It follows that negotiation and establishment of a relationship of mutual trust and respect with an eating-disordered exerciser is possible, in spite of the profound ambivalence that often characterizes eating disorders, provided that fitness professionals and exercise leaders remain faithful to their function as health promoters.

I shall now discuss some practical recommendations that can apply in sports and fitness, and that can be used by all fitness leaders and fitness enthusiasts. Case history 12, reported in Chapter 7, is taken here as an example of good care.

A good action plan

One of the central messages of this book is that the presence of the eating-disordered exerciser should never be ignored, and the exerciser at risk should be warned of the risks and monitored with particular care, to ensure that preventable harm is avoided. This is not only a way of protecting ourselves, as fitness leaders, from legal liability, but also a way of protecting the client's welfare and respecting their autonomy.

A good action plan is one that is framed according to the psychological and physical state of the exerciser. As far as fitness instructors are concerned, an action plan can range from straight exclusion without further justification, to attempted negotiation, to establishment of realistic goals, and to specific exercise and health recommendations.

In case history 12 Marie tells us about how she observed the eating-disordered participant, and how she looked compassionately at the person she is. This means that she tried to understand what was tormenting the sufferer, and tried to empathize with her. Those who have extreme behaviors are often troubled persons, whose life has not allowed them to take better care of themselves, or to love themselves more. The eating disorders sufferer delegates self-love to diet, to exercise: s/he can only accept her/himself, provided that s/he diets and exercises more than feels tolerable.

Marie makes a connection between unemployment, lack of social participation and eating disorders. The client seems to have a lot of time to spend at the gym, so the instructor assumes she might be unemployed and that this might trigger eating disorders. The participant is living in a rather degraded social context, apparently she does not have many friends or a partner, and eating disorders could be precipitated by the sense of loneliness, emptiness and lack of self-worth. The sufferer recovers the sense of control over her own body, and with it, the strength and value that she lacks in other areas of her life.

Marie also notices the peculiar way in which the disordered exerciser is losing weight. Again, she observes the client carefully and with concern.

> At some point, all of a sudden, she started losing lots of weight. You see when losing weight is a problem, because all of a sudden the person becomes thin, they lose weight very quickly, and they looked fine before. Their thinness looks unhealthy, they look dry, rather than thin . . . .

The instructor here gives us important cues for identification. Someone who loses weight very quickly and continues to exercise regardless of the state of debilitation could be at risk of having an eating disorder. Marie provides an example of good and empathic observation skills as applied to the fitness context.

Marie then thinks about what she should do, and decides to speak to the line manager. This is another good choice. She resists the temptation to 'just let the sufferer be': instead, she takes control of the situation, by deciding to speak to someone about her concerns. The line manager also shows consideration towards the instructor. The line manager talks about the private issues of this particular client.

One might worry about confidentiality here. As we have seen in Chapter 8, as a part of the team, relevant information about a client, in appropriate circumstances, might be shared. All colleagues who receive confidential information about a client are bound by confidentiality not to divulge that information. In order to avoid any possible risk of breaching confidentiality, however, it might be prudent to enclose in the initial contract with the client a request to consent to disclosure of information within the team for the purposes of designing proper training programs offering appropriate supervision. The line manager, who in case history 12 is more knowledgeable and has more expertise, takes direct responsibility for the disordered client.

The way the line manager handles the problem is also illustrative of good care. She does not suspend the activities abruptly, but talks to the client and explains the risks in a way that she can understand. For example, she is not talking about risks of mitral valve prolapse, electrolyte imbalances or bone density mineral loss. "You are going too far . . .", is all she says. Moreover, she avoids, at this first stage, worrying or intimidating the client. She attempts to establish a relationship of mutual respect by proposing to work together at this. She accepts her in the gym, provided that she stops dieting. She only asks for suspension of activities when it is clear that the client is not maintaining a safe threshold of body fat, and wishes to see her again in the sports center once she has regained some weight.

Unfortunately, the client seems severely affected by eating disorders, and continues a cycle between recovery and relapse, and Marie cannot say how the story ended. It is possible that an easier liaison with healthcare professionals, as suggested in Chapter 6, might have made a difference in this case. Moreover, as suggested in Chapter 9, the line manager could have tried to persuade the client to seek medical or psychological help, and provided her with some options, such as a helpline number.

Still, case history 12 depicts good care and good communication between parties. Here the professional with more relevant expertise takes the lead, addresses the problem openly, and attempts a negotiation with the client. She shows understanding and a desire to help but does not take risks that would be unprofessional to accept.

Fitness professionals can have great influence over eating-disordered exercisers (see Chapter 6). It is thus important that coaching takes into account the physical as well as psychological vulnerability of these clients. Shangold[1] has offered a number of practical suggestions as to how to help the eating-disordered sufferer with exercising, to gain a better relationship with her/his body. These can also form part of a good action plan.

1   Give basic information to the client. If appropriate, give the client a friendly brochure that explains guidelines for exercise recommended by the American College of Sports Medicine;

2 *Be a role model.* It is not enough to tell a client that she should eat more, when it is evident that we miss our meals, or that she should exercise responsibly when we either do not exercise ourselves or we over-train;[2]

3 *Be credible.* If we are a role model, we show we believe what we say and the client will be more likely to trust us;

4 *Know the clients' needs and goals.* Talk to the client, try to understand what they look for in exercise. A client with eating disorders will be mainly motivated by the idea of keeping low body weight. If you are concerned, for example, about bone thinning, slowly introduce exercise to prevent osteoporosis, and explain why you do so;

5 *Discover the client's preferences.* Exercise can be enjoyable, even if eating-disordered clients live under an overwhelming anxiety of gaining weight. Finding an activity that they truly enjoy can help them to understand that fitness can be fun, and not just a 'moral duty' or a weight-containment measure;

6 *Counter the female stereotype.* Show them that being fit has got nothing to do with being emaciated or an exercise fanatic. Help them to understand that people's value does not reside in their physical shape;

7 *Be persistent.* Working with eating-disordered clients is a long-term challenge;

8 *Learn about the client's life.* Understanding how particular situations and emotions affect the person's exercise routine can help to assist the client to achieve better overall satisfaction.

This advice is here specifically for fitness professionals. It is this category of fitness leaders that face special challenges. However, this action plan can be extended and adapted to all exercise contexts, even those that involve no professional or qualified instructors. An amateur who organizes occasional recreational activities does not have the same legal and professional responsibilities towards their 'guests' as fitness professionals have towards their clients. Yet, s/he has a duty of care, which is both legal and moral, and which is proportionate to her/his competencies and skills. Duty of care, in fact, in tort law,[3] is a legal duty that is imposed *upon any person*, to act towards others with reasonable care, prudence and watchfulness, in order to prevent avoidable harm. If the arguments of this book are acceptable, it follows that turning a blind eye to a disordered exerciser is never a good or ethical option. Talking to someone we trust, or getting medical advice, or discussing the matter with the person afflicted, are all avenues that should be considered by exercise leaders, whether professionals or amateur. Whereas the nature, scope and limits of our duty of care towards others are debatable, what is certain is that it is not and it cannot be ethically acceptable to bring about a preventable accident, and expose someone to severe injury or even death, because of unwillingness to confront openly what is an evident problem. While amateurs should not give training advice to sufferers, some of the key points highlighted above stand for them as well: they must inform participants of the potential risks for health and life, if they have that information; they should be role models; they should be credible; and they should avoid preventable harm. Those who organize events have some degree of moral and legal responsibility for what happens to the participants. It is one of the most general and uncontroversial principles of morality that we should avoid preventable harm to others from occurring, especially when doing so causes minimal or no discomfort to ourselves. There is no ethical justification for standing passively while someone risks a collapse, a fracture, or even death.

Key points: Action plan

- Be a role model.
- Observe your clients.
- Do not ignore what seems to you to be a problem.
- If you manage a gym, ensure that a well-qualified studio coordinator or line manager is available for advice to all staff.
- When you talk to the client, make what you say understandable to her/him.
- At first, do not worry the client if this is not necessary.
- Try to cooperate with the client by negotiating a training program that is acceptable to you both.
- If you request suspension of activities, make sure the client knows this is a temporary situation and you wish her/him to return.

How to deal with our feelings

Approaching someone suspected of having eating disorders feels awkward. Some fitness and sports leaders might be led to believe that, since eating disorders are a delicate problem beyond the remits of their competencies, they should do nothing. Requesting suspension of activities, or even talking about issues that the exerciser does not want to disclose, might feel like poking our noses into people's private lives. Whereas these feelings are understandable, they are not good grounds for ethical decision-making. Ignoring a serious problem just because one feels bad about intervening in it is a self-indulgence that cannot form the basis for moral action. When we are unable to handle a situation by ourselves, we should talk to line managers, more senior or more experienced colleagues, or studio coordinators, and seek advice. Ignoring a serious threat to the exerciser's health, whatever that threat might be, falls short of the professional's duty of care.

Fitness and eating disorders

Due to the confluence of eating-disordered people and other body-abusers in the sports environment, exercise is sometimes mistakenly associated with psychopathology. The argument of this book disconnects regular exercise with psychopathology, and reminds us that regular exercise is necessary at all ages for health and well-being. It has to be accepted that in some cases for the eating disorders sufferer exercise can be harmful and even fatal, and sometimes it might be necessary to request discontinuation of physical activity.

This does not need to be justified with reasons of beneficence. Requesting suspension of activities might not benefit the sufferer. As suggested in Chapter 9, requesting discontinuation of physical activity can be either beneficial or harmful to the client, and there is no precise way of establishing in advance the impact that such a request will have in the short or in the long term. However, this request has to be done at times, because a proper understanding of our function, as fitness leaders, cannot include sales of harm.

In spite of all this, exercise must not be demonized. Arguments that exercise can be inherently harmful or addictive are devoid of any solid ground. Physical activity can have a role in the care of eating disorders. Some rehabilitation centers incorporate exercise in the program of treatment of eating disorders (see Chapter 6). Exercise can increase self-esteem, improve body satisfaction, and change the relationship of the sufferer with their own body. Exercise can help people to love their body, and thus to love themselves. This transition, however, requires especially skilled professionals, well aware of the psychological as well as physical sequels associated with eating disorders.[4]

A new category of fitness professionals

At all levels of education, fitness professionals should receive at least basic information on eating disorders, including their legal responsibilities, and the hazards of exercise for this category of clients.

In addition to this, this book has called on fitness bodies, such as the ACSM, the YMCA or the Register of Exercise Professionals, to create modules that enable specialists to work competently in this area. Professionals should be encouraged to liaise with other professionals and healthcare professionals, a recommendation already found in some US Codes of Ethics (NCSF IV.2 and NIHS II.3, see Chapter 8).

It could be objected that data on the long-term risks and benefits of exercise for eating-disordered users, and data on their physical and psychological response to exercise are not available. Absent empirical data, it is impossible to educate trainers in this area – so the argument might go. It could also be objected that there are no reported lawsuits by eating-disordered exercisers (see Chapter 8, Section 2), and no reported cases of harm or death suffered by eating-disordered exercisers in the sports and fitness context, or as a result of exercise.[5] Education on the legal duties and rights of fitness professionals is thus somehow misplaced.

There are a number of responses to these legitimate concerns: first, we know that up to 20 percent of those with eating disorders die prematurely. And we know that nearly 50 percent of women with eating disorders exercise to excess (see Chapter 1). We might have trained an eating-disordered client who has died. While there has been no lawsuit against a fitness professional either from or on behalf of an eating-disordered exerciser, fitness professionals surely do not want to contribute to bringing about injury or even death. We therefore need to be aware of the potential risks and benefits of exercise and of our legal responsibilities. Second, the fact that a tragedy of this sort has not yet been reported is not a good reason for not thinking about how to avoid it. Moreover, the effects of abnormal eating are known, as well as appropriate exercises for some of these effects. For example, a woman who has had no period for several years is likely to have osteoporosis or osteopenia. Exercise programs should prioritize weight-bearing activities (see Chapter 6). A regular vomiter will be discouraged from taking intense cardio-vascular exercise. These are all types of information that can easily be delivered at various stages of fitness and sports education, even in absence of clear empirical data on the impact of exercise on eating disorders sufferers.

Advice for all

The methods of analysis employed in this book are the methods of applied ethics in general. This book has also brought to the surface the way in which philo-sophical and ethical analysis can be employed to examine the experience of professionals and exercisers at large, to yield practical policy recommendations. The following box summarizes the advice that can be given to various categories of professionals, gym users and exercise enthusiasts.

Box C.1

Advice for studio instructors

- Studio instructors have discretion over who has access to their sessions.
- They are entitled to take their own initiative over supervising a particular client.
- Studio instructors are often unable to monitor the exertion of each client, and therefore they should be particularly cautious with clients who might be at risk.
- Studio instructors should always be able to liaise with line managers, studio coordinators or gym managers over delicate issues, and decide an action plan with them.

Advice for gym instructors

- Gym instructors are generally better able to monitor the activity of eating-disordered exercisers, as they can follow up clients on an individual basis.
- Gym instructors and personal trainers can perform one-to-one assessments, and estimate more carefully the risks for each participant, and the level of exertion of the eating-disordered exerciser.

- Gym instructors should be able to request immediate suspension of activities in case of imminent danger to health and safety of the individual concerned or of others, or in case of extreme, indecent or aggressive behaviors.
- Gym instructors should liaise with the gym manager, and the gym manager might request suspension of activities, and eventual refund of prepaid memberships, for eating-disordered exercisers.

Advice for personal trainers

- Personal trainers have more discretion over whom they will accept as their client, as they are employed by clients on a one-to-one basis.

Advice for gym managers

- Gym managers might decide to take on the responsibility of dealing with the disordered exerciser.
- Gym managers can get involved directly with their clients, and can request suspension of activities and refund unused installments, where appropriate.

Advice for all professionals and exercise leaders

- A client being emaciated, or affected by a severe eating disorder, are adequate grounds for requesting suspension of activities, or for not accepting the participant in the activity in the first place.
- Duty of care requires that, in extreme cases, the client should not be allowed to take even supervised exercise.
- Fitness and sports professionals can request medical certification from the client.
- It should be noted that medical certification, declaring that the exerciser is in sufficiently good health to participate in fitness activities, will cover some professional liability, but it is likely not to protect the instructor from all claims.

Advice for eating-disordered exercisers

- Talk to an instructor you might trust, if you feel you might have an eating problem.
- You do not necessarily need to suspend physical activities: these, however, have to be tailored for you. It is essential you disclose any perils associated with abnormal eating, even if they seem trivial to you.
- Exercise can help you to develop a more balanced attitude towards diet and self: it can help you to achieve better body satisfaction, and thus reduce the severity of eating disorders.

- Normal eating improves physical performance and the effects of exercise on your body, thus normal eating will improve your body satisfaction.
- Make sure your trainer is informed about your condition, and is well qualified to give you nutritional as well as training advice.

Advice for all fitness enthusiasts

- If you are concerned that one of your fitness friends has an eating disorder, you can, in a confidential manner, talk to the fitness leader about your concerns. Sharing your worries is not defamatory and is not a breach of confidentiality.
- If a friend discloses personal information to you in an informal way, you are in principle bound by confidentiality not to divulge that information, even if this is in the best interests of your friend or fellow exerciser. You should instead try to persuade your friend to talk to an instructor about her condition.

Conclusions

Profound ethical issues can often be found where one would not necessarily expect to find them: sports and fitness is one terrain in which many would not presume to face deep ethical quandaries. Fitness professionals get qualified without necessarily having awareness of how many ethical and legal dilemmas they will find in their career. Yet, we have discussed the many ethical and legal predicaments that eating disorders, in particular, generate in the sports and fitness arena.

This book has provided all those involved in fitness and sports with essential information about eating disorders and exercise. Understanding eating disorders and the way exercise can be beneficial or harmful to the sufferer, and interpreting eating disorders, is essential to the purposes of this book, and is essential to ethical sports and fitness practice. Refining the sensitivity towards the drama of vulnerable exercisers is crucial in order to handle satisfactorily the delicate relationship with eating-disordered exercisers.

I have attempted to bring to light the ethical and legal dilemmas that arise, when people with eating disorders in particular, and more generally people who use exercise to the detriment of their health, participate in supervised sports and exercise. I have not given a general answer as to whether eating-disordered exercisers should be allowed – or prevented from taking – supervised exercise. I have disentangled the possible ways of thinking about eating disorders and exercise, and I have shown the strengths and weaknesses of various ways of reasoning about them. Eating disorders have been analyzed in detail, and the way exercise leaders might respond to the predicaments raised by eating disorders has been discussed. I have explored the strengths and weaknesses of various ways of resolving these dilemmas, and proposed policy recommendations that gesture towards a more informed and reasoned way of dealing with the presence of eating-disordered exercisers in the sports and fitness arena.

Notes

Introduction

1 Available at: http://www.nytimes.com/1994/07/28/obituaries/christy-henrich-22-gymnast-who-suffered-from-anorexia.html (accessed 18 November 2009).

2 Available at: http://www.bostonphoenix.com/archive/dance/97/08/GUENTHER.html (accessed 18 November 2009).

3 According to: http://www.b-eat.co.uk/Home (accessed 18 November 2009).

4 Rosalyn Griffiths and Janice Russell, 'Compulsory treatment for anorexia nervosa patients', in Pierre J. V. Beumont and Walter Vandereycken (eds), *Treating Eating Disorders: ethical, legal and personal issues*, New York: New York University Press, 1998, p. 127; Stephen Zipfel, Bernd Löwe and Wolfgang Herzog, 'Medical complications', in Janet Treasure, Ulrike Schmidt and Eric van Furth (eds), *Handbook of Eating Disorders*, 2nd edn, Chichester: Wiley, 2003, ch. 10, pp. 169–190, p. 195; J. Treasure, 'Anorexia and bulimia nervosa', in G. Stein and G. Wilkinson (eds), *Seminars in General Adult Psychiatry*, London: Royal College of Psychiatrists, 1998, pp. 858–902; S. Crow, B. Praus and P. Thuras, 'Mortality from eating disorders: a 5 to 10 years record linkage study', *International Journal of Eating Disorders* 26, 1, 1999, 97–102; D. B. Herzog, D. N. Greenwood, D. J. Doer, A. T. Flores, E. R. Ekeblad, A. Richards, M. A. Blais and M. B. Keller, 'Mortality in eating disorders: a descriptive study', *International Journal of Eating Disorders* 28, 1, 2000, 20–26; Søren Nielsen and Núria Bará Carril, 'Family burden of care and social consequences', in Janet Treasure, Ulrike Schmidt and Eric van Furth (eds), *Handbook of Eating Disorders*, ch. 11, pp. 191–206. Walter Vandereycken suggests that mortality is more often associated with anorexia nervosa than with bulimia nervosa. He also points out that mortality rates are exaggerated because they are based on clinical samples, that is, on a negative selection (personal communication).

5 On 29 July 2009, the *Daily Mail* in the UK reported the case of a girl suffering from anorexia who intended to publicly denounce a well-known gym in the UK for allowing her to join while she was clearly anorexic. See http://www.dailymail.co.uk/femail/article-1202636/Gym-let-join-I-anorexic-step-treadmill-killed-me.html (accessed 18 November 2009).

6 D. W. Mickley, 'Medical dangers of anorexia nervosa and bulimia nervosa', in R. Lemberg and L. Cohn (eds), *Eating Disorders: a reference sourcebook*, Phoenix, AZ: Oryx Press, 1999, p. 46; Jack L. Katz, 'Eating disorders', in M.M. Shangold and G. Mirkin (eds), *Women and Exercise, Physiology and Sports Medicine*, Philadelphia, PA: Davis Company, 1994, pp. 292–312.

7 Fitness professionals include fitness instructors – also called teachers, coaches and trainers – and gym managers and coordinators, who do not always actively teach and who are not always qualified in exercise. Definition adapted from R.A. Thompson and S.R. Trattner, 'Helping athletes with eating disorders', Champaign, IL: Human

Kinetics Publishers, 1993, p. 26. All these categories of people normally share some of the ethical and legal responsibilities towards eating-disordered exercisers. The legal issues analyzed in this book mainly concern professionals. The ethical issues, however, embrace and concern all those involved in physical activity, including amateurs, who might take the lead in non-professional exercise. I shall mainly focus on the relationship between *the professional* and *the eating-disordered client*, but what is said in this book can be extended to non-professional contexts as well.

8　This book refers to supervised exercise, in that it explores the ethical complexities that arise in the particular relationship between exercise leader and eating-disordered participant. A further question is whether people with eating disorders should be allowed to exercise at all. This can be, partly, a matter for clinical judgment. Each individual therapist should thus make a judgment as to whether exercise could be dangerous for the patient or not, and what types of exercise are indicated for each sufferer. This is therefore in a sense a clinical question that has to be answered through a case-by-case assessment. In another sense, however, the question also has ethical dimensions to it. Suppose the eating disorder specialist advises her anorexic patient not to exercise, but the patient continues to do so by herself, in the face of professional advice. Would it be ethical for her to be physically restricted, for her own good? The only way to ensure a person does not exercise by themselves is to implement strict control over their life. There is thus a question of ethical legitimacy of restriction of movement. Anorexics are indeed sometimes physically restricted, for example through coercive hospitalization and treatment. Their movement is severely scrutinized. The question as to whether eating-disordered exercisers should be allowed to take *any* exercise raises thus further issues of whether people's freedom to move, and to harm themselves, be it through dieting, lack of exercise, overeating, and so on, *when they do not involve any third party*, should be restricted. In principle there is limited ethical justification for significant interference with a person's freedom, when they are able to understand the consequence of their behavior, and do not harm or even involve others. The analysis of these issues goes beyond the remit of this book.

9　I use the terms 'moral philosophy' and 'ethics' largely as synonymous. There is no standard accepted differentiation between moral philosophy and ethics. Moral philosophy is generally understood as a branch of philosophy that studies the notions of good and bad, and right and wrong. Moral philosophy includes meta-ethics, which studies the meaning of ethical norms and concepts; normative ethics, which attempts to produce norms of conduct justifiable on reason or on other grounds; and applied ethics, which addresses practical issues such as abortion, euthanasia, distribution of health care resources and others. I tend to privilege the terms 'morality' and 'moral philosophy' because the term 'ethics' is often related to 'ethical codes' and 'ethical principles' or an 'ethos'. Whenever I am not explicitly referring to ethical principles or codes of conduct, I will use the word 'moral'.

10　As I see it.

11　According to the USA bureau census, the US resident population at the time of writing is 307,015,106. According to the National Institute of Diabetes and Kidney Diseases, about one-third of the population is medically obese. This is how I have derived this number. See http://www.census.gov/population/www/popclockus.html for up-to-date census projection and http://www.win.niddk.nih.gov/STATISTICS/ for data on obesity (both accessed 18 November 2009).

12　Simona Giordano, 'Should we force the obese to diet?', editorial, *Journal of Medical Ethics* 34, 5, 2008: 319.

1 Eating disorders: symptoms and facts

1   Binge-eating disorder, for example, is provisionally included both in the DSM-IV (and DSM-IV-TR) and in the ICD-10. However, it is disputed whether it differs sufficiently from bulimia nervosa as to justify as a separate clinical category. See Charles P. Bull, 'Binge Eating Disorder: diagnostic issues', *Current Opinion in Psychiatry* 17, 1, 2004: 43–48.

2   William Gull, 'Anorexia Nervosa', *The Lancet* 1, 1888, 516–17.

3   Kindly provided by Walter Vandereycken.

4   M. Selvini Palazzoli, *Anoressia mentale, dalla terapia individuale alla terapia familiare*, Milano: Feltrinelli, 1998 (originally 1963), p. 46, my translation (English version: *Self-starvation: from the intrapsychic to the transpersonal approach to anorexia nervosa*, London: Human Context Books/Chaucer, 1974); see also H. Bruch, *The Golden Cage: the enigma of anorexia nervosa*, West Compton House: Open Books, 1980, p. 4, my emphasis.

5   Bruch, *The Golden Cage*, p. 90.

6   *ICD-10*, F10–19.

7   A brief account of the classification of eating disorders may be found in Paul E. Garfinkel, 'Classification and diagnosis of eating disorders', in Christopher G. Fairburn and Kelly D. Brownell (eds), *Eating Disorders and Obesity*, 2nd edn, London: The Guilford Press, 2002, Ch. 28, pp. 155–61.

8   Available at: http://en.wikipedia.org/wiki/Obesity (accessed 23 November 2009).

9   H. Bruch, *Eating Disorders: obesity, anorexia nervosa and the person within*, London: Routledge, 1974.

10 World Health Organization (WHO), *ICD-10 Classification of Mental and Behavioural Disorders: clinical descriptions and diagnostic guidelines*, Geneva: WHO, 1992; see also American Psychiatric Association (APA), *Diagnostic and Statistical Manual of Mental Disorders, DSM-IV TR (Text Revision)*, 4th edn, Washington DC: APA, 2000, 307.1.

11 Nathan J. Smith, 'Excessive weight loss and food aversion in athletes simulating anorexia nervosa', *Pediatrics* 66, 1, July 1980, 139–42.

12 M.T. Pugliese, F. Lifshitz, G. Grad, P. Fort and M. Marks-Katz, 'Fear of obesity: a cause of short stature and delayed puberty', *The New England Journal of Medicine* 309, 1983, 513–18.

13 William D. McArdle, Frank I. Katch and Victor L. Katch, *Sports and Exercise Nutrition*, Philadelphia, PA: Lippincott Williams and Wilkins, 1999, p. 469.

14 Linda Smolak and Michael P. Levine, *Body Image in Children*, Washington: American Psychological Association, 2001.

15 Both Sarah's and Jude's stories are fictional but based on real narratives.

16 Anne Ward and Simon Gowers, 'Attachment and childhood development', in Janet Treasure, Ulrike Schmidt and Eric van Furth (eds), *Handbook of Eating Disorders*, 2nd edn, Chichester: Wiley, 2003, Ch. 6, pp. 103–20, p.115.

17 P. J. McKenna, 'Disorders with overvalued ideas', *British Journal of Psychiatry* 145, 1984, 579–85; A. Freeman and V. B. Grenwood (eds), *Cognitive Therapy: an overview, in cognitive therapy, application in psychiatry and medical settings*, New York: Human Science, 1987. For a full account of the cognitive distortions of people with eating disorders see Simona Giordano, *Understanding Eating Disorders*, Oxford: Oxford University Press, 2005, Ch. 12.

18 Bruch, *The Golden Cage*.

19 H. Draper, 'Anorexia nervosa and respecting a refusal of life-prolonging therapy: a limited justification', *Bioethics* 14, 2, 2000, 120–33.

20 Bruch, *Eating Disorders*, Ch. 4.

21 Simona Giordano, 'Choosing death in cases of anorexia nervosa: should we ever let people die from anorexia?', in Charles Tandy (ed), *Death and Anti-Death*, volume 5, Palo Alto: CA: Ria University Press, 2007, Ch. 9.

22 See testimonial and discussion on BBC Radio 4, *Inside the Ethics Committee*, 5 September 2007. Available at: http://www.bbc.co.uk/radio4/science/ethicscommittee_20070905.shtml (accessed 23 November 2009).

23 Giordano, *Understanding Eating Disorders*.

24 *ICD-10*. In the ICD-9 'Anorexia nervosa' is reported at 307.1, together with 'Other and unspecified disorders of eating', at 307.5.

25 *DSM-IV-TR*, 307.1; Bruch, *The Golden Cage*, pp. 72–90; see also K.A. Halmi, 'Current concepts and definitions', in G. Szmukler and C. Dare (eds), *Handbook of Eating Disorders*, Chichester: John Wiley & Sons, 1995, pp. 29–44.

26 William D. McArdle, Frank I. Katch and Victor L. Katch, *Sports and Exercise Nutrition*, p. 470; Jonathan Mond, Tricia Cook Myers, Ross Crosby, Phillipa Hay and James Mitchell, '"Excessive exercise" and eating-disordered behaviour in young adult women: further evidence from a primary care sample', *European Eating Disorders Review* 16, 2008: 215–21; Caroline Davis and Simone Kaptein, 'Anorexia nervosa with excessive exercise: a phenotype with close links to obsessive-compulsive disorder', *Psychiatric Research* 142, 2006: 209–17. The notion of 'excessive' exercise is obviously open to interpretation. Moreover, what is excessive to one individual might not be to another. For discussion of the notion of 'excessive exercise' see J.M. Mond, P.J. Hay, B. Rogers, C. Owen and P.J.V. Beumont, 'Relationship between exercise behaviour, eating-disordered behaviour and quality of life in a community sample of women: when is exercise "excessive"?', *European Eating Disorders Review*, 12, 4, 2004: 265–72. See also J.M. Mond, P.J. Hay, B. Rodgers and C. Owen, 'An update on the definition of "excessive exercise" in eating disorders research', *International Journal of Eating Disorders* 39, 2, 2005: 147–53; J.M. Mond and R.M. Calogero, 'Excessive exercise in eating disorder patients and in healthy women', *Australian and New Zealand Journal of Psychiatry*, 43, 3, 2009: 227–34. In this book exercise is 'excessive' when the individual suffers physical harm as a consequence of the amount, duration and frequency of exercise. See Chapter 6 for symptoms of excessive exercise.

27 R.L. Horne, 'Disturbed body image in patients with eating disorders', *American Journal of Psychiatry* 148, 2, 1991: 211–15.

28 R. Shafran and C.G. Fairburn, 'A new ecologically valid method to assess body size estimation and body size dissatisfaction', *International Journal of Eating Disorders* 32, 2002: 458–65.

29 P.K. Bowden *et al.*, 'Distorting patient or distorting instrument? Body shape disturbance in patients with anorexia nervosa and bulimia', *British Journal of Psychiatry* 155, 1989: 196–201; G. Noordenbos, 'Differences between treated and untreated patients with anorexia nervosa', *British Review of Bulimia and Anorexia Nervosa* 3, 2, 1989: 55–60; M.J. Tovée, P.J. Benson, J.L. Emery, S.M. Mason and E.M. Cohen-Tovée, 'Measurement of body size and shape perception in eating-disordered and control observers using body-shape software', *British Journal of Psychology* 94, 4, 2003: 501–16; R.M. Gardner and E.D. Bokenkamp, 'The role of sensory and non-sensory factors in body size estimations of eating disorder subjects', *Journal of Clinical Psychology* 52, 1, 1996: 3–15. For further discussion see Rick M. Gardner and Cathy Moncrieff, 'Body image distortion in anorexics as a non-sensory phenomenon: a signal detection approach', *Journal of Clinical Psychology* 44, 2, 2006: 101–107; R. Rodgers and H. Chabrol, 'Parental attitudes, body image disturbance and disordered eating amongst adolescents and young adults: a review', *European Eating Disorders Review*, 17, 2, 2009: 137–51; S.D. Christman, M. Bentle, C.L. Niebauer, 'Handedness differences in body image distortion and eating disorder symptomatology', *European Eating Disorders Review* 40, 3, 2007: 247–56; S. Vocks, T. Legenbauer, H. Ruddel and N.F. Troje, 'Static and dynamic body image in bulimia nervosa: mental representation of body dimensions and biological motion patterns', *International Journal of Eating Disorders* 40, 1, 2007: 59–66; L.K. George Hsu and Theresa A. Sobkiewicz,

'Body image disturbance: time to abandon the concept for eating disorders?', *International Journal of Rating Disorders*, 10, 1, 2006: 15–30.

30 Clare Farrell, Michelle Lee and Roz Sharfan, 'Assessment of body size estimation: a review', *European Eating Disorders Review* 13, 2005: 74–88, p. 77.

31 As reported by Walter Vandereycken, 'Denial of illness in anorexia nervosa – a conceptual review: part 1, Diagnostic significance and assessment', *European Eating Disorders Review* 14, 2006, 341–51, p. 346. See also Walter Vandereycken, 'Denial of illness in anorexia nervosa – a conceptual review: part 2, Different forms and meanings', *European Eating Disorders Review* 14, 2006, 352–68.

32 Mental Health Act 2007, The Stationery Office Limited, London. For an account of the amendments to the Mental Health Act 1983, see http://www.opsi.gov.uk/acts/acts2007/en/ukpgaen_20070012_en.pdf (accessed 23 November 2009).

33 Clinical Resource Efficiency Support Team (CREST), established under the auspices of the Central Medical Advisory Committee in 1988. See 2004 guidelines at http://www.crestni.org.uk/tube-feeding-guidelines.pdf (accessed 23 November 2009).

34 For an early study, see R.A. Vigersky (ed), *Anorexia Nervosa*, London: Raven Press, 1977; A.E. Andersen, *Practical Comprehensive Treatment of Anorexia Nervosa and Bulimia*, London: Edward Arnold, 1985; George Szmukler, Chris Dare and Janet Treasure, *Handbook of Eating Disorders: theory, treatment and research*, Chichester: Wiley, 1995. A more recent discussion can be found in S. Abraham, *Eating Disorders: the facts*, 6th edn, Oxford: Oxford University Press, 2008. See also Riccardo Dalla Grave, Simona Calugi and Giulio Marchesini, 'Compulsive exercise to control shape or weight in eating disorders: prevalence, associated features and treatment outcome', *Comprehensive Psychiatry* 49, 2008: 346–52; C. Laird Birmingham, Stephen Touyz and Jane Harbottle, 'Are anorexia nervosa and bulimia nervosa separate disorders? Challenging the "transdiagnostic" theory of eating disorders', *European Eating Disorders Review* 17, 2009: 2–13.

35 For a detailed account, see Andersen, *Practical Comprehensive Treatment*; see also C.G. Fairburn and G.T. Wilson, *Binge Eating*, New York: Guilford Press, 1993. See also Giordano, *Understanding Eating Disorders*, Ch. 1.

36 Richard Gordon, *Anoressia e bulimia, anatomia di un'epidemia sociale*, Milano: Raffaello Cortina, 1991, p. 96. The original English text is Richard Gordon, *Eating Disorders: analysis of a social epidemic*, Oxford: Blackwell, 1990, 2nd edn, 2000.

37 O.W. Hill, 'Epidemiological aspects of anorexia nervosa', *Advances in Psychosomatic Medicine* 9, 1977: 48–62.

38 This information was originally published at this site: http://www.edauk.com/sub_some_statistics.htm. In 2009, the *Independent* reported similar figures (http://www.independent.co.uk/life-style/health-and-families/healthy-living/one-million-britons-suffer-from-eating-disorders—so-why-is-so-little-being-done-to-help-1642775.html (accessed 2 December 2009).

39 See http://www.childline.org.uk/ (accessed 23 November 2009).

40 Rosalyn Griffiths and Janice Russell, 'Compulsory treatment for anorexia nervosa patients', in Walter Vandereycken and Pierre J.V. Beumont (eds), *Treating Eating Disorders: ethical, legal and personal issues*, New York: New York University Press, 1998, p. 127.

41 Marlene Busko, 'High suicide rate in anorexia linked to lethal methods, not fragile health', *Journal of Affective Disorders* 107, 2008: 231–36.

42 *DSM-IV-TR*, 2000, 307.1; 370.51.

43 *ICD-10*, F 50.

44 Hans Wijbrand Hoek, 'Distribution of eating disorders', in Fairburn and Brownell (eds), *Eating Disorders*, Ch. 41, pp. 233–37, p. 233.

45 By *prevalence* is meant the number of all cases of eating disorders in the population at a particular moment or during a specified period; by *incidence* is meant the number

of *new cases* of eating disorders per year in the population. Hoek, 'Distribution of eating disorders'.

46 Mervat Nasser, *Culture and Weight Consciousness*, London and New York: Routledge, 1997, p. 12.

47 As quoted in Mervat Nasser, *Culture and Weight Consciousness*, p. 61.

48 An account of the effects of eating disorders in males may be found in Arnold E. Andersen, 'Eating disorders in males', in Fairburn and Brownell (eds), *Eating Disorders*, Ch. 33, pp. 188–92, p.188. See also Antonia Baum, 'Eating disorders in the male athlete', *Sports Medicine* 36, 1, 2006: 1–6.

49 Manfred Fichter and Heidelinde Krenn, 'Eating disorders in males', in Treasure, Schmidt and van Furth (eds), *Handbook of Eating Disorders*, Ch. 23, pp. 369–83.

50 Elena Faccio, *Il disturbo alimentare, modelli ricerche e terapie*, Roma: Carocci, 1999, p. 30, 32; Fichter and Krenn, 'Eating disorders', pp. 369–83.

51 I have previously published these cases in Giordano, *Understanding Eating Disorders*, Ch. 7.

52 Mara Selvini Palazzoli, S. Cirillo, M. Selvini and A.M. Sorrentino, *Ragazze anoressiche e bulimiche, la terapia familiare*, Milan: Raffaello Cortina Editore, 1998, pp. 61–62; my emphasis and translation.

53 See for example N.S. Witton, P. Leichner, P. Sandhu-Sahota and D. Filippelli, 'The Research Directions Survey: patient and parent perspectives on eating disorder research', *Eating Disorders* 15, 3, May 2007: 205–16; Vesna Vidovic, Vesna Jure, Ivan Begovac, Mirta Mahnik and Gorana Tocilj, 'Perceived family cohesion, adaptability and communication in eating disorders', *European Eating Disorders Review* 13, 1, January 2005: 19–28; Joanna Patching and Jocelyn Lawler, 'Understanding women's experiences of developing an eating disorder and recovering: a life-history approach', *Nursing Inquiry*, 16, 1, March 2009: 10–21; James Lock and Kathleen Fitzpatrick, 'Evidence-based treatments for children and adolescents with eating disorders: family therapy and family-facilitated cognitive-behavioral therapy', *Journal of Contemporary Psychotherapy*, 37, 3, 2007: 145–155.

54 For more recent findings, see Walter Vandereycken, 'Families of patients with eating disorders', in Fairburn and Brownell (eds), *Eating Disorders*, Ch. 38.

55 Bruch, *Eating Disorders*, p. 262.

56 Bruch, *Eating Disorders*, p. 264, my emphasis.

57 Jenna Whitney, Rachel Haigh, John Weinman and Janet Treasure, 'Caring for people with eating disorders: factors associated with psychological distress and negative caregiving appraisals in carers of people with eating disorders', *British Journal of Clinical Psychology* 46, 3, November 2007: 413–29.

58 Mervat Nasser and Melanie Katzman, 'Sociocultural theories of eating disorders: an evaluation in thought', in Janet Treasure, Ulrike Schmidt and Eric van Furth (eds), *Handbook of Eating Disorders*, pp. 139–150, p. 139; Solange Cook-Darzens, Catherine Doyen, Bruno Falissard and Marie-Christine Mouren, 'Self-perceived family functioning in 40 French families of anorexic adolescents: implications for therapy', *European Eating Disorders Review* 13, 4, July 2005: 223–36; Paul Rhodes, Andrew Baillie, Jac Brown and Sloane Madden, 'Parental efficacy in the family-based treatment of anorexia: preliminary development of the Parents Versus Anorexia Scale (PVA)', *European Eating Disorders Review* 13, 6, November 2005: 399–405.

59 Selvini Palazzoli, Cirillo, Selvini and Sorrentino, *Ragazze anoressiche*, pp. 22–23 and p. 122.

60 Jane Ogden, *The Psychology of Eating: from healthy to disordered behaviour*, Oxford: Blackwell, 2003, p. 92.

61 Marilyn Duker and Roger Slade, *Anorexia and Bulimia: how to help*, 2nd edn, Buckingham and Philadelphia: Open University Press, 2003, p. 124; Richard Gordon, *Anoressia e bulimia*, p. 96. In the English version see Ch. 6.

62 Gordon, *Anoressia e bulimia*, p. 96. In the English version see Ch. 6.

63 Jan B. Lackstrom and D. Blake Woodside, 'Families, therapists and family therapy in eating disorders', in Vandereycken and Beumont (eds), *Treating Eating Disorders*, pp. 106–126, p. 107.

64 Lackstrom and Woodside, 'Families, therapists and family therapy', p. 108.

65 Simona Giordano, 'Persecutors or victims? The moral logic at the heart of eating disorders', *Health Care Analysis*, 11, 3, 2003: 219–28.

66 See for example the Palazzoli team at http://www.centrimaraselvini.it/home.php (accessed 23 November 2009).

67 Lies Depestele and Walter Vandereycken, 'Families around the table: experiences with a Multi-Family Approach in the treatment of eating disordered adolescents', *International Journal of Child Health and Adolescent Health* 2, 2009: 255–61.

68 J.F. Fitzgerald and R.C. Lane, 'The role of the father in anorexia', *Journal of Contemporary Psychotherapy* 30, 1, 2000: 71–84.

69 B.S. Turner, *The Body and Society: explorations in social theory*, Oxford: Blackwell, 1984, p. 192.

70 Duker and Slade, *Anorexia and Bulimia*, p. 127.

71 Anne Ward and Simon Gowers, 'Attachment and childhood development', in Treasure, Schmidt and van Furth (eds), *Handbook of Eating Disorders*, Ch. 6, pp. 103–20, p. 115.

72 Ivan Eisler, Daniel Le Grange and Eia Asen, 'Family interventions', in Treasure, Schmidt and van Furth (eds), *Handbook of Eating Disorders*, Ch. 18, pp. 291–310, pp. 291, 292.

73 Richard Gordon provides an analysis of eating disorders as an ethnic condition in *Eating Disorders*, Ch. 1; Pierre Beumont and Walter Vandereycken, 'Challenges and risks for health care professionals', in Vandereycken and Beumont (eds), *Treating Eating Disorders*, Ch. 1, p. 1.

74 Vandereycken and Beumont (eds), *Treating Eating Disorders*, Ch. 3.

75 Selvini Palazzoli, Cirillo, Selvini and Sorrentino, *Ragazze anoressiche e bulimiche*, p. 201. See also L.C. Joyce, 'Meanings of eating disorders discerned from family treatment and its implications for family education: the case of Shenzhen', *Child and Family Social Work* 12, 4, November 2007: 409–16. Although Mervat Nasser reports that the phenomenon is emerging in other societies and cultures. She has offered an interesting and rich comparative analysis of prevalence and incidence of eating disorders in different countries, and among different races. See Nasser, *Culture*, Ch. 2.

76 Nasser, *Culture*, p. 14. Nasser gives other examples of culture-bound syndromes: *amox*, a homicidal tendency found in south-east Asia; *latha*, a trance-like state in the south Pacific; and *koro*, the delusion of receding penis.

77 Nasser, *Culture*, p. 106.

78 This box was previously published in my *Understanding Eating Disorders*, Ch. 8.

79 Selvini Palazzoli, Cirillo, Selvini and Sorrentino, *Ragazze anoressiche*, p. 201; Gordon, *Eating Disorders*, Ch. 3.

80 Mervat Nasser and Melanie Katzman, 'Sociocultural theories of eating disorders: an evaluation in thought', in Treasure, Schmidt and van Furth (eds), *Handbook of Eating Disorders*, Ch. 8, p. 145.

81 Among the most influential, I should mention Hilde Bruch, 'Four decades of eating disorders', in D.M. Gardner and P.E. Garfinkel (eds), *Handbook for the Psychotherapy of Anorexia Nervosa and Bulimia*, New York: Guilford Press, 1985, p. 9.

82 M. MacSween, *Corpi Anoressici*, Milano: Feltrinelli, 1999, p. 39. The English original version is *Anorexic Bodies: a feminist and social perspective*, London: Routledge, 1995. See Ch. 2.4. See also Mary Briody Mahowald, 'To be or not be a woman: anorexia nervosa, normative gender roles, and feminism', *The Journal of Medicine and Philosophy* 17, 1992: 233–51.

83 MacSween, *Corpi Anoressici*, p. 71. English version, Ch. 3.3.

84　A.H. Crisp, 'Diagnosis and outcome of anorexia nervosa: the St George's View', *Proceedings of the Royal Society of Medicine* 70, 1977: 464–70.

85　S. Orbach, '*Hunger Strike: the anorexic's struggle as a metaphor for our age*', 2nd edn, London: Karnac Books, 2005; see also Mahowald, 'To be or not to be a woman'.

86　Selvini Palazzoli, *Anoressia mentale*, p. 75. This situation has been described in the subsequent literature in nearly identical terms. See for example B. Dolan, 'Why women? Gender issues and eating disorders, Introduction', in B. Dolan and I. Gitzinger (eds), *Why Women? Gender issues and eating disorders*, London: The Athlone Press, 1994; Nasser, *Culture*, p. 65.

87　Nasser, *Culture*, p. 63.

88　MacSween, *Corpi anoressici*, p. 39. English version see Ch. 5.

89　Gordon, *Eating Disorders*, Ch. 5.

90　Ogden, *The Psychology of Eating*, p. 72.

91　Gordon, *Eating Disorders*, Ch. 5.

92　*The Guardian*, 31 May 2000, available at: http://www.guardian.co.uk/uk/2000/may/31/audreygillan (accessed 23 November 2009).

93　Essential fat is the fat stored in the bone marrow, the heart, lungs, liver, spleen, kidneys, intestines, muscles and lipid-rich tissues of the nervous system. Essential fat includes sex-specific fat, in females. This is thought to include deposits in the pelvic, buttock and thigh regions.

94　F.S. Parekh and R.A. Schmidt, 'In pursuit of an identity: fashion marketing and the development of eating disorders', *British Food Journal* 105, 4–5, 2003: 220–38.

95　Selvini Palazzoli, Cirillo, Selvini and Sorrentino, *Ragazze anoressiche*, p. 215; Orbach, *Hunger Strike*.

96　Christine Halse, Anne Honey and Desiree Boughtwood, 'The paradox of virtue: (re)thinking deviance, anorexia and schooling', *Gender and Education* 19, 2, March 2007: 219–35, p. 228.

97　Halse, Honey and Boughtwood, 'The paradox of virtue', p. 225.

98　Halse, Honey and Boughtwood, 'The paradox of virtue', p. 225.

99　Walter Vandereycken and Ron Van Deth, *From Fasting Saints to Anorexic Girls: the history of self-starvation*, London: Athlone Press, 1994.

100　Giordano, *Understanding Eating Disorders*, Chs 8 and 10.

2　The effects of abnormal eating

1　I wish to thank Walter Vandereycken particularly for checking on the medical facts and data contained in this chapter and correcting me whenever appropriate.

2　Rosalyn Griffiths and Janice Russell, 'Compulsory treatment for anorexia nervosa patients', in Walter Vandereycken and Pierre J. V. Beumont (eds), *Treating Eating Disorders: ethical, legal and personal issues*, New York: New York University Press, 1998, p. 127.

3　Stephen Zipfel, Bernd Löwe and Wolfgang Herzog, 'Medical complications', in Janet Treasure, Ulrike Schmidt and Eric van Furth (eds), *Handbook of Eating Disorders*, 2nd edn, Chichester: Wiley, 2003, Ch. 10, pp. 169–190, p. 195. As also explained in Chapter 1, some, for example Walter Vandereycken, consider these statistics as portraying a too negative picture of eating disorders, as they are based on clinical negative samples, and are therefore biased (personal communication). However, probably all other statistics are too. So, they are, in principle, at least as reliable as other similar statistics.

4　A.E. Taylor *et al.*, 'Impact of binge eating on metabolic and leptin dynamics in normal young women', *Journal of Clinical and Endocrinological Metabolism* 84, 2, 1999: 428–34. Some researchers suggest that factors other than body weight may affect leptin changes. See P. Monteleone, A. Di Lieto, A. Tortorella, N. Longobardi and M. Maj, 'Circulating leptin in patients with anorexia nervosa, bulimia nervosa or binge-eating disorder: relationship to body weight, eating patterns, psychopathology and endocrine changes', *Psychiatric Research* 94, 2, 2000: 121–29. A more technical account of endocrine abnormalities, electrolyte and metabolic abnormalities; cardiac and cardiovascular disturbances and gastrointestinal complications may be found in Katherine A. Halmi, 'Physiology of anorexia nervosa and bulimia nervosa', in Christopher G. Fairburn and Kelly D. Brownell (eds), *Eating Disorders and Obesity*, 2nd edn, London: The Guilford Press, 2002, Ch. 48, pp. 267–71; and Claire Pomeroy and James E. Mitchell, 'Medical complications of anorexia nervosa and bulimia nervosa', in Fairburn and Brownell (eds), *Eating Disorders*, Ch. 50, pp. 278–83. This chapter contains information on dermatological abnormalities, dental complications and immunological disorders associated with eating disorders.

5   Suzanne F. Abraham, Bianca Pettigrew, Catherine Boyd and Janice Russell, 'Predictors of functional and exercise amenorrhea among eating- and exercise-disordered patients', *Human Reproduction* 21, 1, 2006: 257–61.

6   A.P. Winston, A.E.F. Alwazeer and M.J.G. Bankart, 'Screening for osteoporosis in anorexia nervosa: prevalence and predictors of reduced bone mineral density', *International Journal of Eating Disorders* 41, 3, 2008: 284–87.

7   David L. Nichols, Charlotte F. Sanborn and Eve V. Essery, 'Bone density and young athletic women: an update', *Sports Medicine* 37, 11, 2007: 1001–14; K.K. Miller, E.E. Lee, E.A. Lawson, M. Misra, J. Minohan, S.K. Grinspoon, S. Gleysteen, D. Mickley, D. Herzog and A. Klibanski, 'Determinants of skeletal loss and recovery in anorexia nervosa', *Journal of Clinical Endocrinology and Metabolism* 91, 8, 2006: 2931–37.

8   P.S. Mehler, A.L. Sabel, T. Watson and A.E. Andersen, 'High risk of osteoporosis in male patients with eating disorders', *International Journal of Eating Disorders* 41, 7, 2008: 666–72.

9   Jack L. Katz, 'Eating disorders', in M.M. Shangold and G. Mirkin (eds), *Women and Exercise, Physiology and Sports Medicine*, Philadelphia: Davis Company, 1994, pp. 292–312.

10 R. Shelley, *Anorexics on Anorexia*, London: Jessica Kingsley Publishers, 1992, p. 7; U. Cuntz, G. Frank, P. Lehnert and M. Fichter, 'Interrelationships between the size of the pancreas and the weight of patients with eating disorders', *International Journal of Eating Disorders* 27, 2000: 297–303; A. Marcos, 'Eating disorders: a situation of malnutrition with peculiar changes in immune system', *European Journal of Clinical Nutrition* 54, 2000, Suppl. 1: 61–64.

11 D.W. Mickley, 'Medical dangers of anorexia nervosa and bulimia nervosa', in R. Lemberg and L. Cohn (eds), *Eating Disorders: a reference sourcebook*, Phoenix: Oryx Press, 1999, p. 47; see also P.S. Mehler and T.D. MacKenzie, 'Treatment of osteopenia and osteoporosis in anorexia nervosa: a systematic review of the literature', *International Journal of Eating Disorders* 42, 3, 2009: 195–201.

12 Lyn Patrick, 'Eating disorders: a review of the literature with emphasis on medical complications and clinical nutrition', *Alternative Medicine Review*, June 2002: 184–207.

13 Y. Jayasinghe, S.R. Grover and M. Zacharin, 'Current concepts in bone and reproductive health in adolescents with anorexia nervosa', *International Journal of Obstetrics and Gynaecology*, 115, 3, 2008: 304–15.

14 I owe this observation to Walter Vandereycken. Personal communication. See also A.D. Divasta, T.J. Beck, M.A. Petit, H.A. Feldman, M.S. LeBoff and C.M. Gordon, 'Bone cross-sectional geometry in adolescents and young women with anorexia nervosa: a hip structural analysis study', *Osteoporosis International* 18, 6, 2007: 797–804; J. Dominguez, L. Goodman, S.S. Gupta, L. Mayer, S.F. Etu, B.T. Walsh, J. Wang, R. Pierson and M.P. Warren, 'Treatment of anorexia nervosa is associated with increases in bone mineral density, and recovery is a biphasic process involving both nutrition and return of menses', *American Journal of Clinical Nutrition* 86, 1, 2007: 92–99.

15 The mitral valve is the left atrioventricular valve. The prolapse consists of the bulging or billowing of one or more parts of the mitral valve towards the left atrium and, depending on its degree and intensity, it may cause the blood to flow back to the left atrium. This prolapse is found in otherwise healthy subjects and in most cases sports are not contraindicated. However, in particularly serious cases the mitral valve prolapse may cause cardiac insufficiency. It is therefore important to check the degree of the anomaly and whether any cardiac anomaly is associated with it.

16 Mickley, 'Medical dangers of anorexia nervosa and bulimia nervosa', p. 47; Katz, 'Eating disorders', p. 299.

17 Adenosine triphosphate, or ATP, is one of the carriers of energy in human cells. It is often described as the 'coin' or the 'currency' of the cell.

18 Katz, 'Eating disorders', p. 300.

19 Mickley, 'Medical dangers of anorexia nervosa and bulimia nervosa', p. 47.

20 Katz, 'Eating disorders', p. 299.

21 C.M. Shisslak and M. Crago, 'Eating disorders among athletes', in R. Lemberg and L. Cohn (eds), *Eating Disorders: a reference sourcebook*, Phoenix: Oryx Press, 1999, pp.79–83, p. 81.

22 A. Kiss, H. Bergmann, T.A. Abatzi, C. Schneider, S. Wiesnagrotzki, J. Höbart, G. Steiner-Mittelbach, G. Gaupmann, A. Kugi and G. Stacher-Janotta, 'Oesophageal and gastric motor activity in patients with bulimia nervosa', *Gut* 31, 3, 1990: 259–65; Donald T. Fullerton, Susan Neff and Carl J. Getto, 'Persistent functional vomiting', *International Journal of Eating Disorders* 12, 2, 1992: 229–33.

23 Patrick, 'Eating disorders'.

3 Biological and clinical explanations of eating disorders

1 This chapter partly relies on previous research, published in Simona Giordano, *Understanding Eating Disorders*, Oxford: Oxford University Press, 2005, Chapter 4.

2 See for example S.P. Grossman, 'Contemporary problems concerning our understanding of brain mechanisms that regulate food intake and body weight', in A.J. Stunkard and E. Stellar (eds), *Eating Disorders*, New York: Raven Press, 1984, pp. 5–15; J. Treasure and A. Holland, 'Genetic factors in eating disorders', in G.I. Szmukler and C. Daren (eds), *Handbook of Eating Disorders*, Chichester: John Wiley & Sons, 1995, pp. 49–65.

3 A. Kipman, L. Bruins-Slot, C. Boni, N. Hanoun, J. Adès, P. Blot, M. Hamon, M.C. Mouren-Siméoni and P. Gorwood, '5-HT2A gene promoter polymorphism as a modifying rather than a vulnerability factor in anorexia nervosa', *European Psychiatry* 17, 2002: 227–29, p. 229; D.E. Grice, K.A. Halmi, M.M. Fichter, M. Strober, D.B. Woodside, J.T. Treasure, A.S. Kaplan, P.J. Magistretti, D. Goldman, C.M. Bulik, W.H. Kaye and W.H. Berrettini, 'Evidence for a susceptibility gene for anorexia nervosa on Chromosome 1', *American Journal of Human Genetics* 70, 2002: 787–92.

4 Elizabeth Winchester and David Collier, 'Genetic aetiology of eating disorders and obesity', in Janet Treasure, Ulrike Schmidt and Eric van Furth (eds), *Handbook of Eating Disorders*, 2nd edn, Chichester: Wiley, 2003, Ch. 3, pp. 35–64.

5 Winchester and Collier, 'Genetic aetiology of eating disorders and obesity'.

6 Winchester and Collier, 'Genetic aetiology of eating disorders and obesity', p. 36.

7 Lyn Patrick, 'Eating disorders: a review of the literature with emphasis on medical complications and clinical nutrition', *Alternative Medicine Review*, June 2002: 184–207.

8 Elena Faccio, *Il disturbo alimentare, modelli ricerche e terapie*, Roma: Carocci, 1999, p. 92; see also E. Waugh *et al.*, 'Offspring of women with eating disorders', *International Journal of Eating Disorders* 25, 2, 1999: 123–33.

9 Janet Polivy and C. Peter Herman, 'Causes of eating disorders', *Annual Review of Psychology* 53, 2002: 187–213.

10 Lyn Patrick, 'Eating disorders'.

11   Elena Faccio, *Il disturbo alimentare*, p. 92.
12   Michael Strober and Cynthia M. Bulik, 'Genetic epidemiology of eating disorders', in Christopher G. Fairburn and Kelly D. Brownell (eds), *Eating Disorders and Obesity*, 2nd edn, London: The Guilford Press, 2002, Ch. 42, pp. 238–43, p. 239.
13   Strober and Bulik, 'Genetic epidemiology of eating disorders', p. 238.
14   Winchester and Collier, 'Genetic aetiology of eating disorders and obesity', p. 39.
15   According to D.A. Campbell *et al.*, 'Fine mapping of human 5-HTR2 a gene to chromosome 3914 and identification of two highly polymorphic linked markers suitable for association studies in psychiatric disorders', *Genetic Testing* 1, 4, 1997: 297–99. But clinical results are conflicting.
16   D. E. Grice *et al.* 'Evidence', p. 787.
17   Kelly L. Klump, Stephen Wonderlich, Pascale Lehoux, Lisa R.R. Lilenfeld and Cynthia M. Bulik, 'Does environment matter? A review of nonshared environment and eating disorders', *International Journal of Eating Disorders* 31, 2002: 118–135; L.R. Lilenfeld, W.H. Kaye, C.G. Greeno, K.R. Merikangas, K. Plotnicov, C. Pollice, R. Rao, M. Strober, C.M. Bulik and L. Nagy, 'A controlled family study of anorexia nervosa and bulimia nervosa: psychiatric disorders in first-degree relatives and effects of proband comorbidity', *Archives of General Psychiatry* 55, 1998: 603–610; M. Strober, R. Freeman, C. Lampert, J. Diamond and W. Kaye, 'Controlled family study of anorexia and bulimia nervosa: evidence of shared liability and transmission of partial syndromes', *American Journal of Psychiatry*, 157, 2000: 393–401; Margarita C.T. Slof-Op 't Landt, Eric F. van Furth, Ingrid Meulenbelt, Eline P. Slagboom, Meike Bartels, Dorret I. Boomsma and Cynthia M. Bulik, 'Eating disorders: from twin studies to candidate genes and beyond', *Twin Research and Human Genetics* 8, 5, October 2005: 467–482.
18   A. Holland *et al.*, 'Anorexia nervosa: a study of 34 pairs of twins and one set of triplets', *British Journal of Psychiatry* 145, 1984: 414–19.
19   T. Wade *et al.*, 'Genetic and environmental risk factors for the weight and shape concerns characteristic of bulimia nervosa', *Psychological Medicine* 28, 4, 1998: 761–77; Colin A. Ross, 'Overestimates of the genetic contribution to eating disorders', *Ethical Human Psychology and Psychiatry* 8, 2, 2006: 123–33.
20   Kelly L. Klump *et al.*, 'Does environment matter?', p. 120.
21   Genetics influence all of one's biological functions, including those of the nervous system. It is therefore important to note that the research on the central nervous system is not at odds with research on the genetics of eating disorders, but can be placed in a wider panorama of research on the biological or organic causes of eating disorders.
22   Martina de Zwaan, 'Basic neuroscience and scanning', in Treasure, Schmidt and van Furth (eds), *Handbook of Eating Disorders*, Ch. 5, pp. 89–101; A very articulated account of the neurotransmitter activity in anorexia and bulimia nervosa may be found in Walter H. Kaye, 'Central nervous system neurotransmitter activity in anorexia nervosa and bulimia nervosa', in Fairburn and Brownell, *Eating Disorders and Obesity*, Ch. 49, pp. 272–77.
23   D.T. Fullerton *et al.*, 'Plasma immunoreactive beta-endorphin in bulimics', *Psychological Medicine* 16, 1986: 59–63.
24   J. Morley, A.S. Levine and D. D. Krahn, 'Neurotransmitter regulation of appetite and eating', in B.J. Blinder, B.F. Chaitin and R.S. Goldstein (eds), *Eating Disorders: medical and psychological bases of diagnosis and treatment*, PMA, New York, 1988, pp. 11–19. Serotonin also seems to be involved. See D.C. Jimerson *et al.*, 'Low serotonin and dopamine metabolite concentrations in cerebrospinal fluid from bulimic patients with frequent binge episodes', *Archives of General Psychiatry* 49, 2, 1992: 132–38, quoted in B.J. Blinder, M.C. Blinder and V.A. Sanathara, 'Eating disorders and addiction', *Psychiatric Times* 15, 12, 1998: 30–34.
25   Blinder, Chaitin and Goldstein (eds), *Eating Disorders*, pp. 11–19.

26 J.P. Reneric and M.P. Bouvard, 'Opioid receptor in antagonists in psychiatry: beyond drug addiction', *Drugs* 10, 5, 1998: 365–82; M.A. Marazzi *et al.*, 'Endogenous codeine and morphine in anorexia and bulimia nervosa', *Life Sciences* 60, 20, 1997: 1741–47; see also M.A. Marazzi *et al.*, 'Male/female comparison of morphine effect on food intake: relation to anorexia nervosa', *Biochemistry and Behavior* 53, 2, 1998: 433–35; Anna Capasso, Claudio Putrella and Walter Milano, 'Recent clinical aspects of eating disorders', *Reviews on Recent Clinical Trials* 4, 1, 2009: 63–69.

27 A connection between opioid receptors and the release of dopamine associated with feeding has been suggested by M.T. Taber *et al.*, 'Opioid receptor modulation of feeding-evoked dopamine release in the rat nucleus accumbens', *Brain Research* 785, 1, 1998: 24–30, quoted in Blinder *et al.*, 'Eating disorders and addiction'.

28 A.S. Kaplan and P.E. Garfinkel, 'The neuroendocrinology of anorexia nervosa', in R. Collu, G.M. Brown and R. Van Loon (eds), *Clinical Neuroendocrinology*, London: Blackwell, 1988, p. 117.

29 W. Kaye *et al.*, 'Serotonin neuronal function and selective serotonin reuptake inhibitor treatment in anorexia and bulimia nervosa', *Biological Psychiatry* 44, 9, 1998: 825–38.

30 P. Dally, J. Gomez and A.J. Isaacs, *Anorexia nervosa*, London: Heinemann, 1979, p. 198. See also Polivy and Herman, 'Causes of eating disorders'.

31 Dally, Gomez and Isaacs, *Anorexia nervosa*, p. 161.

32 Kaplan and Garfinkel, 'The neuroendocrinology of anorexia nervosa', p. 188.

33 Dally, Gomez and Isaacs, *Anorexia Nervosa*, p. 172.

34 Patrick, 'Eating disorders'.

35 Kaplan and Garfinkel, 'The neuroendocrinology of anorexia nervosa', p. 109.

36 Patrick, 'Eating disorders', p. 109.

37 For a study on the growth hormones, cortisol and prolactin, in the specific case of bulimia, see D. Goldbloom *et al.*, 'The hormonal response to intravenous 5-hydroxytryptophan in bulimia nervosa', *Psychosomatic Medicine* 52, 2, 1990: 225–26.

38 M. Pawlikowski and J. Zarzycki, 'Does the impairment of the hypothalamic-pituitary-gonadal axis in anorexia nervosa depend on increased sensitivity to endogenous melatonin?', *Medical Hypotheses* 52, 2, 1999: 111–13.

39 Kaplan and Garfinkel, 'The neuroendocrinology of anorexia nervosa', pp. 107, 200.

40 Dally, Gomez and Isaacs, *Anorexia nervosa*, p. 27.

41 Dally, Gomez and Isaacs, *Anorexia nervosa*, p. 35.

4 Eating disorders, exercise and addiction

1 A picture of Gus Goose is available at http://coa.inducks.org/character.php?c=GU (accessed 24 November 2009).

2 B.J. Blinder, M.C. Blinder and V.A. Sanathara, 'Eating disorders and addiction', *Psychiatric Times* 15, 12, 1998: 30–34.

3 E.J. Cumella, Z. Kally and A.D. Wall, 'Treatment responses of inpatient eating disorder women with and without co-occurring obsessive-compulsive disorder', *Eating Disorders* 15, 2007: 111–24.

4 A.M. Bastiani, M. Altemus, T.A. Pigott, C. Rubenstein, T.E. Weltzin and W.H. Kaye, 'Comparison of obsessions and compulsions in patients with anorexia nervosa and obsessive compulsive disorder', *Biological Psychiatry* 39, 1996: 966–69; T.A. Fahy, A. Osacar and I. Marks, 'History of eating disorders in female patients with obsessive compulsive disorder', *International Journal of Eating Disorders* 14, 1993: 439–43. Some research regards exercise as compulsive for eating disorders. See Riccardo Dalle Grave, Simona Calugi and Giulio Marchesini, 'Compulsive exercise to control shape or weight in eating disorders: prevalence, associated features, and treatment outcome', *Comprehensive Psychiatry* 49, 2008: 346–52; R. Finzi-Dottan and E. Zubery,

'The role of depression and anxiety in impulsive and obsessive-compulsive behaviors among anorexic and bulimic patients', *Eating Disorders* 17, 2009: 162–82.

5 J. M. Darley, S. Glucksberg, L.J. Kamin and R.A. Kinchla, *Psychology*, London and Englewood Cliffs: Prentice Hall, 1984, pp. 141–52, pp. 498–503. C. Landau, 'Substance abuse', in *Encyclopedia of Psychology*, New York: John Wiley & Sons, 1984, pp. 382–83.

6 Available at: http://library.adoption.com/parenting-girls/for-parents-and-caregivers-compulsive-exercise/article/6656/1.html (accessed 24 November 2009).

7 Darley *et al.*, *Psychology*, pp. 141–52, 498–503.

8 I owe this further observation to Walter Vandereycken.

9 I admit that this sentence is inaccurate. It in fact suggests that the chemical reactions come chronologically and logically *before* pleasure. But it is possible that my son produces these chemical reactions *because* he gives me pleasure (for example, because I love him, because I have bonded with him or because a biological instinct useful to the preservation of the species makes me attached to my baby). The pleasure thus does not follow, either chronologically or logically, the chemical reactions but these might be interdependent events. For the purposes of this book, this is not relevant and therefore will be not discussed further. Moreover, my distinction between the different senses in which we talk of addiction is not exhaustive, but highlighting the difference between these two meanings of the word *addiction*, used both in ordinary language and in clinical literature, is essential to understanding what happens to people with eating disorders and whether they can reasonably be considered as affected by a form of addiction.

10 Kristin M. Von Ranson and Stephanie E. Cassin, 'Eating disorders and addiction: theory and evidence', in Jerome S. Rubin (ed), *Eating Disorders and Weight Loss Research*, New York: Nova Science Publishers, 2007, pp. 1–37, p. 12.

11 I owe this observation to Michael McNamee.

12 The philosophical debate on freedom of the will is several centuries long and cannot be discussed here. However, those who are interested in the conflict between different volitions, in the devotion and how they shape our sense of self, might want to read Harry Frankfurt, and in particular, 'Freedom of the will and the concept of the Person', *Journal of Philosophy* 68, 1, 197: 5–20; see also Harry Frankfurt, *Necessity, Volition and Love*, Cambridge: Cambridge University Press, 1988, in particular pages 129–44.

13 In philosophy, those who believe that, in spite of these determining factors, human beings are free, are called *compatibilists*; those who believe that, in the presence of these determining factors, human beings are not free are called *incompatibilists*. These hold that either a) there are no determining factors and therefore human beings are free or b) there are determining factors and therefore human beings are not free. This work assumes a compatibilist position.

14 World Health Organization (WHO), *ICD-10 Classification of Mental and Behavioural Disorders: clinical descriptions and diagnostic guidelines*, Geneva: WHO, 1992.

15 Blinder *et al.*, 'Eating disorders'.

16 P.F. Sullivan, C.M. Bulik, J.L. Fear and A. Pickering, 'Outcome of anorexia nervosa', *American Journal of Psychiatry* 155, 7, 1998: 939–46.

17 Elena Faccio, *Il disturbo alimentare, modelli ricerche e terapie*, Rome: Carocci, 1999, pp. 69, 44.

18 C.C. Holderness, J. Brooks-Gunn, and M.P. Warren, 'Comorbidity of eating disorders and substance abuse review of literature', *International Journal of Eating Disorders* 16, 1, 1994: 1–34; see also C.M. Grilo, M.A. White and R.M. Masheb, 'DSM-IV psychiatric disorder comorbidity and its correlates in binge eating disorder', *International Journal of Eating Disorders* 42, 2009: 228–34; J. Jordan, P.R. Joyce, F.A. Carter, J. Horn, V.V.W. McIntosh, S.E. Luty, J.M. McKenzie, C.M.A. Frampton, R.T. Mulder and C.M. Bulik, 'Specific and nonspecific comorbidity in anorexia

nervosa', *International Journal of Eating Disorders* 41, 2008: 47–56; F. Fernandez-Aranda, A.P. Pinheiro, L.M. Thornton, W.H. Berrettini, S. Crow, M.M. Fichter, K.A. Halmi, A.S. Kaplan, P. Keel, J. Mitchell, A. Rotondo, M. Strober, D.B. Woodside, W.H. Kaye and C.M. Bulik, 'Impulse control disorders in women with eating disorders', *Psychiatry Research* 157, 2008: 145–57: J.M. Swinbourne and S.W. Touyz, 'The co-morbidity of eating disorders and anxiety disorders: a review', *European Eating Disorders Review* 15, 2007: 253–74.

19 Von Ranson and Cassin, 'Eating disorders and addiction', pp. 10–11.

20 P. De Silva and S. Eysenck, 'Personality and addictiveness in anorexic and bulimic patients', *Personality and Individual Differences* 8, 5, 1987: 749–51.

21 See, for example, R.A. Vigersky (ed) *Anorexia Nervosa*, London: Raven Press, 1977; A.E. Andersen, *Practical Comprehensive Treatment of Anorexia Nervosa and Bulimia*, London: Edward Arnold, 1985; George Szmukler, Chris Dare and Janet Treasure, *Handbook of Eating Disorders: theory, treatment and research*, Chichester: Wiley, 1995; S. Abraham, *Eating Disorders: the facts*, 6th edn, Oxford: Oxford University Press, 2008.

22 For a detailed account, see A.E. Andersen, *Practical Comprehensive Treatment*; see also C.G. Fairburn and G.T. Wilson, *Binge Eating*, New York: Guilford Press, 1993. The person with eating disorders is not concerned about putting on 'too much' fat or about excessive weight. Normally people with eating disorders have a fear of fat and weight even if they are severely underweight.

23 G. Razzoli, *La bulimia nervosa*, Milan: Sonzogno, 1995, p. 62.

24 T. Tuomisto *et al.*, 'Psychological and physiological characteristics of sweet food addiction', *International Journal of Eating Disorders* 25, 2, 1999: 139–75.

25 E. Faccio, *Il disturbo alimentare*, p. 69.

26 E. Faccio, *Il disturbo alimentare*, p. 69.

27 G. Terence Wilson, 'Eating disorders and addictive disorders', in Christopher G. Fairburn and Kelly D. Brownell (eds), *Eating Disorders and Obesity*, 2nd edn, London: Guilford Press, 2002, Ch. 35, pp. 199–203, p. 199.

28 It has to be noticed that tolerance and withdrawal are not essential to a diagnosis of substance use dependence, because only three of the criteria set out by diagnostic manuals have to be met. However, tolerance and withdrawal are important indicators that addiction has a neurophysiological nature, and that it is being used in the strong sense. If it is used in the weak sense, as we have seen above, we are all potentially ill. I therefore regard tolerance and withdrawal as crucially important to a diagnosis of addiction, as, as explained, it only makes sense to regard anorexia, bulimia and over-training as addictions in the strong sense.

29 Marilyn Duker and Roger Slade, *Anorexia Nervosa and Bulimia: how to help*, 2nd edn, Buckingham and Philadelphia: Open University Press, 2003, p. 32.

30 This box is adapted from Rick Kiddle, *Studio Cycling*, London: A&C Black, 2004, p. 137.

31 Von Ranson and Cassin, 'Eating disorders and addiction', p. 19.

32 Jack L. Katz, 'Eating disorders', in Mona M. Shangold and Gabe Mirkin (eds), *Women and Exercise: physiology and sports medicine*, 2nd edn, Philadelphia: Davis Company, 1994, Ch. 17, pp. 292–312, p. 307; C. Davis, D.K. Katzman and C. Kirsh, 'Compulsive physical activity in adolescents with anorexia nervosa: a psychobehavioral spiral of pathology', *The Journal of Nervous and Mental Disease* 187, 6, 1999: 336–42.

33 Benjamin Allegre, Marc Souville, Pierre Therme and Mark Griffiths, 'Definitions and measures of exercise dependence', *Addiction Research and Therapy* 14, 2006: 631–46.

34 Ivan Eisler and Daniel le Grange, 'Excessive exercise and anorexia nervosa', *International Journal of Eating Disorders* 9, 4, 1990: 377–86: see also A. Yates, K. Leehey and C.M. Shisslak, 'Running: an analogue of anorexia?', *New England Journal of Medicine* 308, 1983: 251. Quoted in Katz, 'Eating disorders', p. 307.

35 Simona Giordano, *Understanding Eating Disorders*, Oxford: Oxford University Press, 2005, Chs 5 and 6.

36 Simona Giordano, 'Qu'un souffle de vent . . .', *Medical Humanities* 28, 1, 2002: 3–8; Giordano, *Understanding Eating Disorders*.

37 Giordano, *Understanding Eating Disorders*.

38 Giordano, *Understanding Eating Disorders*.

39 Vincent T. De Vita, Samuel Hellman and Steven A. Rosenberg (eds), *Cancer: principles and practice of oncology*, 7th edn, Philadelphia: Lippincott Williams & Wilkins, 2005; Kai Ming Chan, Mary Anderson and Edith M.C. Lau, 'Exercise interventions: defusing the world's osteoporosis time bomb', *Bulletin of the World Health Organization* 81, 2003: 827–30; Susan L. Woods, Erika Sivarajan Froelicher, Sandra Adams Motzer and Elizabeth J. Bridges (eds), *Cardiac Nursing*, 5th edn, Philadelphia: Lippincott Williams & Wilkins, 2005; K. Amy and Priscilla M. Clarkson, 'The acute effects of exercise and inactivity on vascular function', *Medicine and Science in Sports and Exercise* 39, 5, Suppl. S84, May 2007: 881; WHO, *Active Ageing: a policy framework*, WHO/NMH/NPH/02.8, Geneva: World Health Organization, 2002, p. 28.

40 WHO, *Active Ageing*, p. 14.

41 Available at: http://www.euro.who.int/observatory/toppage (accessed 24 November 2009).

42 See also David W. Kalisch, Tetsuya Aman and Libbie A. Buchele, 'Social and health policies in OECD countries: a survey of current programmes and recent developments', *OECD Working Papers*, Paris: OECD, 1998, 71.

43 WHO, *Life in the 21st Century: A Vision for All* (World Health Report), Geneva: World Health Organization, 1998.

44 Denise E. Wilfley, Carlos M. Grilo and Kelly D. Brownell, 'Exercise and regulation of body weight', in Shangold and Mirkin (eds), *Women and Exercise*, Ch. 2, pp. 27–59, p. 31.

45 Council on Health Promotion, 'Promoting physical activity', *British Columbia Medical Journal* 2, 47, March 2005: 85–86. Available online at: http://www.bcmj.org/promoting-physical-activity (accessed 24 November 2009).

46 WHO, 'Healthy ageing is vital for development', Press Release WHO/24, 9 April 2002.

47 WHO, *Active Ageing*, p. 28; see also WHO, 'Growing older – staying well', pp. 6–7; P. Dargent-Molina, F. Favier, H. Grandjea, C. Badoin, A.M. Schott, E, Hausherr, P.J. Meunier and G. Breart for EPIDOS Group, 'Fall-related factors and risk of hip fracture: the EPIDOS prospective study', *Lancet* 348, 1996: 145–49; C.N. Merz and J.S. Forrester, 'The secondary prevention of coronary heart disease', *American Journal of Medicine* 102, 1997: 573–80; J. Ruuskanen and I. Ruoppila, 'Physical activity and psychological well-being among people aged 65 to 84 years', *Age and Ageing* 24, 1995: 292–96; P.J. O'Connor, L.E. Aenchbacher and R.K. Dishman, 'Physical activity and depression in the elderly', *Journal of Aging and Physical Activity* 1, 1993: 34–58; H. Sugisawa, J. Liang and X. Liu, 'Social network, social support and mortality among older people in Japan', *Journal of Gerontology* 49, 1, 1994: 3–13; Kalisch, Aman and Buchele, 'Social and health policies', p. 114.

48 See for example the Oxford English Dictionary, the definition reported in Section 3.

5 Media and eating disorders

1 Available at: http://www.settelen.com/diana_eating_disorders.htm (accessed 25 November 2009).

2 Available at: http://www.caringonline.com/eatdis/celebrities_p.html (accessed 25 November 2009).

3 British Medical Association, *Eating Disorders, Body Image and the Media*, London: BMA, 2000, p. 43. See also the Newswire article at: http://web2.bma.org.uk/press rel.nsf/wlu/RWIN-4KTDYW?OpenDocument&vw=wfmms (accessed 4 December 2009).

4 Janet L. Treasure, Elizabeth R. Wack and Marion E. Roberts, 'Models as a high-risk group: the health implications of a size zero culture', *British Journal of Psychiatry* 192, 2008: 243–44. Debate available at: http://www.b-eat.co.uk/Publications/Reports, 'Has Fashion got its house in order?' (accessed 11 February 2010).

5 Audrey Gillan, 'Skinny models "send unhealthy message"', *The Guardian*, 31 May 2000, available at: http://www.guardian.co.uk/uk/2000/may/31/audreygillan (accessed 25 November 2009); Libby Brooks, 'Body image debate needs a fuller picture', *The Guardian*, 30 May 2000, available at: http://www.guardian.co.uk/print/0%2C38 58%2C4023718–103699%2C00.htm (accessed 25 November 2009).

6 CCN, 'Skinny models banned from catwalk', 13 September 2006, available at: http://www.cnn.com/2006/WORLD/europe/09/13/spain.models/index.html (accessed 25 November 2009).

7 Vanessa Thorpe and David Smith, 'Fashion week chief rejects catwalk ban on super-thin', *The Guardian*, 17 September 2006, available at: http://www.guardian.co.uk/business/2006/sep/17/marksspencer.uknews (accessed 25 November 2009).

8 Marl Sweney, 'Unilever bans size zero models', *The Guardian*, 8 May 2007, at http://www.guardian.co.uk/media/2007/may/08/advertising.unilever (accessed 25 November 2009).

9 Luke Harding, 'Ruthlessly exposed', *The Guardian*, 8 April 2005, available at: http://www.guardian.co.uk/world/2005/apr/08/worlddispatch.germany (accessed 25 November 2009).

10 Vanessa Beecroft's website at www.vanessabeecroft.com (accessed 25 November 2009) contains a number of references and links to her exhibitions and interesting reviews.

11 Of course, I am not claiming that my interpretation of her works is the best or the only one possible. One of the captivating elements of art is that it is meant to elicit emotions, which are necessarily subjective, and thus are open to various interpretations. It is possible that other observers will see different messages in Beecroft's exhibitions. It seems to me unequivocal, though, that her artworks are provoking and invariably informed by some type of political or social protest. For example, in her last exhibition at the Biennale of Venice, one of the most important world arts exhibitions, she represented the war in Darfur. She scattered a huge canvas with 'blood' (red paint). Around 30 African women were lying nearly naked, one next to each other, immobile and silent, for hours, like dead bodies thrown in a common grave. In the meantime Beecroft herself was pouring 'blood' on them with her paintbrush. In this exhibition as well, we can read a protest not only against the preventable deaths of many people in this forgotten region of Africa, but also the claim of the responsibility of seemingly passive Western observers who, by omission, share the responsibility for the blood shed. Into her exhibitions of women, I similarly read a protest against the use of the woman in the public eye. See also Nick Johnstone, 'Dare to bare', *The Guardian*, 13 March 2005, available at: http://arts.guardian.co.uk/features/story/0,,1436353,00.html (accessed 25 November 2009).

12 See for example Albrecht Dürer, 'Four books on human proportion', in E.G. Holt (ed.), *A Documentary History of Art*, vol. 1, Princeton: Princeton University Press, 1981.

13 These words are by Supervert, who published a beautiful critique of Beecroft's performances, 'Vanessa Beecroft: Classic Cruelty', *Parkett Magazine*, 1999, republished at: http://supervert.com/essays/art/vanessa_beecroft (accessed 25 November 2009).

14 Of course, the excess of fat also covers individuality, but corpulence, for other reasons explained elsewhere, is not acceptable: it is the sign of indulgence, excessive love for

bodily pleasures, weakness and moral collapse. See Simona Giordano, 'Labia mea domine', in Richard Ashcroft, Angus Dawson, Heather Draper and John McMillan (eds), *Principles of Health Care Ethics*, 2nd edn, Chichester: John Wiley and Sons, 2007, Ch. 61. See also later sections of this chapter.

15 Jonathan M. Mond, Philippa J. Hay, Brian Rodgers, Cathy Owen and Pierre J. V. Beumont, 'Beliefs of women concerning causes and risk factors for bulimia nervosa', *Australian and New Zealand Journal of Psychiatry* 38, 6, 2004: 463.

16 Available at: http://www.b-eat.co.uk/ProfessionalStudentResources/Studentinformation-1/Effectsofthemedia (accessed 25 November 2009).

17 Mervat Nasser, *Culture and Weight Consciousness*, London and New York: Routledge, 1997, p. 57.

18 Nasser, *Culture*, p. 57.

19 Jane Ogden, *The Psychology of Eating: from healthy to disordered behaviour*, Oxford: Blackwell, 2003, p. 90.

20 Maree Burns and Bicola Gavey, '"Healthy weight" at what cost? Bulimia and a discourse of weight control', *Journal of Health Psychology* 9, 4, 2004: 549–65.

21 This raises questions on the relationship between eating disorders and other abuses of exercise, for example, extreme body building, which seem more common in men. In Chapter 7 one case history raises this point. If the arguments of this chapter are accepted, the connection between these various types of abuse of exercise is a certain conception of morality and of what it is to be a worthwhile person. See also Gary S. Goldfield, Dan W. Harper and Arthur G. Blouin, 'Are bodybuilders at risk for an eating disorder?', *Eating Disorders* 6, 1998: 133–57; Lebur Rohman, 'The relationship between anabolic androgenic steroids and muscle dysmorphia: a review', *Eating Disorders* 17, 3, 2009: 187–99; Frederick Grieve, 'A conceptual model of factors contributing to the development of muscle dysmorphia', *Eating Disorders* 15, 1, 2007: 63–80; P.E. Mosley, 'Bigorexia: bodybuilding and muscle dysmorphia', *European Eating Disorders Review* 17, 3, 2009: 191–98; Antonia Baum, 'Eating disorders in the male athlete', *Sports Medicine* 36, 1, 2006: 1–6; H.G. Pope, K.A. Phillips and R. Olivardia, *The Adonis Complex: the secret crisis of male body obsession*, New York: Free Press, 2000.

22 Ogden, *The Psychology of Eating*, p. 71.

23 Richard Gordon, *Anoressia e bulimia, anatomia di un'epidemia sociale*, Milan: Raffaello Cortina, 1991, p. 92. Original English version *Anorexia and Bulimia: anatomy of a social epidemic*, Oxford: Blackwell, 1990, Ch. 5; 2nd edn, 2000.

24 Mark Conner, Charlotte Johnson and Sarah Grogan, 'Gender, sexuality, body image and eating behaviours', *Journal of Health Psychology* 9, 4, 2004: 505–15.

25 Hilde Bruch, *Eating Disorders: obesity, anorexia nervosa and the person within*, London: Routledge and Kegan Paul, 1974; Mara Selvini Palazzoli, *L'anoressia mentale, dalla terapia individuale alla terapia famigliare*, Milan: Feltrinelli, 1963. English translation: *Self-starvation: from the intrapsychic to the transpersonal approach to anorexia nervosa*, London: Human Context Books/Chaucer, 1974.

26 Anna Krugovoy Silver, *Victorian Literature and the Anorexic Body*, Cambridge: Cambridge University Press, 2003, p. 8.

27 The following information is taken from G. Reale and D. Antiseri, *Il pensiero occidentale dalle origini a oggi*, Milan: La Scuola, 1984, Vol. 1.

28 Plato, *Phaedo*, translated by G.M.A. Grube, *Five Dialogues*, Indianapolis: Hackett Publishing, 1981, pp. 93–155, p. 103, quoted in Silver, *Victorian Literature*, p. 171.

29 Aristotle, *De Anima (On the Soul)*, Harmondsworth: Penguin, 1986, iii 4, 429a9–10; cf. iii 3, 428a5; iii 9, 432b26; iii 12, 434b3. Aristotle, *Metaphysics*, London: Penguin, 1998, i 1, 980a21; *De Anima* ii 3, 414b18; iii 3, 429a6–8.

30 The same dichotomy may also be found in Eastern traditions. However, I am not attempting to give an account of the different religious and metaphysical traditions present in the world. I am only trying to understand eating disorders and these

particular aspects of Western culture may help us to understand eating-disordered behavior, regardless of whether similar values may be found in different cultural contexts.

31 G. Ryle, *The Concept of Mind*, London: Penguin, 1978, pp. 13–14.

32 W. Vandereycken and R. van Deth, *From Fasting Saints to Anorexic Girls: the history of self-starvation*, London: Athlone Press, 1994, Chs 2 and 11.

33 'Asceticism', *The Catholic Encyclopedia*, vol. 1, Copyright (c) 1907 by Robert Appleton Company, Online Edition Copyright (c) 2003 by Kevin Knight; available at: http://www.newadvent.org/cathen/01767c.htm (accessed 25 November 2009).

34 Istituto della Enciclopedia Italiana, *Vocabolario della Lingua Italiana*, Rome: Treccani, 1986.

35 Sometimes lightness is associated with immorality (in the epithet 'a light woman', for example). I owe this observation to Harry Lesser. Despite this and other possible negative meanings of lightness, it is often presented as a positive quality to have and the ideals of beauty, spirituality and purity have often and for a long time been associated with lightness. See Giordano, *Understanding Eating Disorders*, Chs 5 and 6.

36 W. Vandereycken and R. ven Deth, *From Fasting Saints*, Chs 2 and 11.

37 *The Catholic Encyclopedia*.

38 With exceptions.

39 Bruch, *Eating Disorders*, p. 25; M. Weber, *The Protestant Ethic and the Spirit of Capitalism*, London: George Allen & Unwin, 1976, p. 166; M. MacSween, *Anorexic Bodies: a feminist and social perspective*, London: Routledge, 1995, p. 211.

40 These claims are partly logical, partly empirical, and partly speculative. Yet it seems to me that it is important to try to understand people's experiences and behavior, especially when we are confronted with the choice of restricting their freedom for their own sake. As I said at the beginning of this book, eating disorders raise the issue of whether we should force people to behave differently from the way they would otherwise, for their own welfare. The underlying belief that runs through this and other works of mine is that in order to understand how one should act in given circumstances, one needs to understand as clearly as possible what is going on. Ethics, I have argued elsewhere, rests on psychology and might on occasions collapse into psychology, in all those cases where, when we analyze a problem, we realize that the ethical questions no longer make sense, and all we need to ask is *why?* (no longer *what should we do?*). Understanding eating disorders in terms of values may be incomplete and partly speculative, but it is still helpful as it is a way of making sense of behaviours and experiences that otherwise would not make sense, and that would be (and have often been) judged as irrational or unintelligible. The use of force and limitations of freedom are often 'justified' on the basis that others' behaviors are irrational and unintelligible. But often this use of paternalism is ethically dubious.

41 M. Duker and R. Slade, *Anorexia Nervosa and Bulimia: how to help*, Buckingham and Philadelphia: Open University Press, 2003, pp. 121–22; Vandereycken and van Deth, *From Fasting Saints*, Chs 2 and 11; M. Lawrence, *The Anorexic Experience*, London: The Women's Press, 1984, pp. 32–35; Bruch, *Eating Disorders*, p. 25, note 27.

42 Weber, *The Protestant Ethic*, p. 180; B.S. Turner, *The Body and Society: explorations in social theory*, London: Blackwell, 1984, Chs 3, 7 and 8.

43 Available at: http://www.shirleys-wellness-cafe.com/fasting.htm (accessed 25 November 2009).

44 Available at: http://www.mindyourbody.info/testimonials.html (accessed 25 November 2009).

45 Available at: http://www.idrocolonterapia.info/idrocolonuk.htm (accessed 25 November 2009).

46 American Cancer Society, 'Colon therapy', available at: http://www.cancer.org/docroot/ETO/content/ETO_5_3x_Colon_Therapy.asp (accessed 25 November 2009).

47 See M. Warin, 'Miasmatic calories and saturating fats: fear of contamination in anorexia', *Culture, Medicine and Psychiatry* 27, 1, 2003: 77–93.

48 MacSween, *Anorexic Bodies*, pp. 211, 217–18. On the dichotomy of full/empty see Ch. 7.

49 A similar psychological dynamic seems to be found in other groups who commit self-harm. Renée Kyle argues that self-harm is a search for autonomy, power and control over the self. R.L. Kyle, 'Harming and healing: young women and the development of the autonomous self', PhD thesis, Faculty of Arts, University of Wollongong, Australia, 2006, available at: http://ro.uow.edu.au/theses/599 (accessed 25 November 2009).

50 Nasser, *Culture*, pp. 58 ff.

6 Exercise and eating disorders

1 From the BBC news archive for 9 November 1993, available at: http://news.bbc.co.uk/onthisday/hi/dates/stories/november/9/newsid_2515000/2515739.stm (accessed 4 December 2009).

2 D.E. Wilfley, C.M. Grilo and K.D. Brownell, 'Exercise and regulation of body weight', in Mona M. Shangold and Gabe Mirkin (eds), *Women and Exercise, Physiology and Sports Medicine*, 2nd edn, Philadelphia: Davis Company, 1994, pp. 33–34.

3 As pointed out in the Introduction to this book, I only consider structured and supervised exercise because the issue of whether people should be allowed to exercise by themselves (going for a run, for example) raises additional ethical questions regarding the legitimacy of intrusion on people's freedom of movement, which cannot be analyzed here. Obvious cases of restrictions of freedom to exercise concern people with anorexia who are hospitalized for their condition. When anorexics are hospitalized they are normally prevented from taking unsupervised exercise. Some report doing sit-ups at night or jogging in their bedroom when nobody sees them. Anorexic inpatients are normally prevented from burning out or throwing up as a part of their re-feeding program. The analysis of the ethics of these invasions of people's freedom goes beyond the remits of this book.

4 R.A. Thompson, S.R. Trattner, *Helping Athletes with Eating Disorders*, Champaign, IL: Human Kinetics Publishers, 1993, p. 26; I. Eisler and D. Le Grange, 'Excessive exercise and anorexia nervosa', *International Journal of Eating Disorders* 9, 1990: 377–86; A. Yates, C. Shisslack, M. Crago, *et al.*, 'Overcommitment to sport: is there a relationship to eating disorders?' *Clinical Journal of Sports Medicine* 4, 1994: 39–46.

5 Morc Coulson, *The Fitness Instructor's Handbook: a complete guide to health and fitness*, London: A&C Black, 2007, p. 91.

6 Edward T. Howley and B. Don Franks, *Fitness Professional's Handbook*, 5th edn, London: Human Kinetics, 2007, p. 446.

7 Pyruvate is an enzyme which assists in the utilization of a phosphate for the production of ATP.

8 Neil Armstrong and Joanne Welsman, *Young People and Physical Activity*, Oxford: Oxford University Press, 1997, p. 23.

9 Mary L. O'Toole and Pamela S. Douglas, 'Fitness: definition and development', in Shangold and Mirkin (eds), *Women and Exercise*.

10 One of the first studies on the benefits of exercise on cardiovascular capacity and body composition was published in the 1940s by Cureton, Balke and Karpovich, quoted by Howley and Franks, *The Fitness Professional's Handbook*, p. 4.

11 I owe this observation to Michael McNamee.

12 T.K. Cureton, *Physical Fitness Appraisal and Guidance*, St. Louis: Mosby, 1947; O'Toole and Douglas, 'Fitness', p. 3.

13 As quoted in Coulson, *The Fitness Instructor's Handbook*, p. 151.

14 As quoted in Coulson, *The Fitness Instructor's Handbook*, p. 151.

15 Some of these definitions are adapted from O'Toole and Douglas, 'Fitness', p. 4; others are from Coulson, *The Fitness Instructor's Handbook*, pp. 151–57.

16 Adapted from Coulson, *The Fitness Instructor's Handbook*, p. 157.

17 Howley and Franks, *Fitness Professional's Handbook*, p. 8.

18 For a fuller account see O'Toole and Douglas, 'Fitness', p. 4.

19 Howley and Franks, *Fitness Professional's Handbook*, p. 8.

20 Liv Duesund and Finn Skarderud, 'Use the body and forget the body: training anorexia nervosa with adapted physical activity', *Clinical Child Psychology and Psychiatry* 8, 1, 2003: 53–72.

21 Proprioceptors are sensory receptors found in the body (in the muscles, tendons, joints and inner ear).

22 Morag Close suggested that proprioceptors might be compromised in children with anorexia nervosa. See Morag Close, 'Physiotherapy and exercise', in Bryan Lask and Rachel Bryant-Waugh (eds), *Anorexia Nervosa and Related Eating Disorders in Childhood and Adolescence*, 2nd edn, Hove: Psychology Press, 2000, pp. 289–305, p. 295.

23 O'Toole and Douglas, 'Fitness', p. 18; see also D. Corrado, F. Migliore, C. Bassoand and G. Thiene, 'Exercise and the risk of sudden cardiac death', *Herz* 31, 6, 2006: 553–58.

24 C. Davis, D.K. Katzman, S. Kaptein, *et al.* 'The prevalence of high level exercise in the eating disorders: etiological implications', *Comprehensive Psychiatry* 38, 1997: 321–26.

25 Wilfley, Grilo and Brownell, 'Exercise and regulation of body weight', pp. 46–47.

26 G. Egger, N. Champion and A. Bolton, *The Fitness Leader's Handbook*, London: A&C Black, 2002, pp. 14 and 18.

27 Jack L. Katz, 'Eating disorders', in Shangold and Mirkin (eds), *Women and Exercise*, pp. 292–312.

28 J.A. Pruitt, R.V. Kappius and P.S. Imm, 'Sports, exercise, and eating disorders', in L. Diamant (ed), *Psychology of Sports, Exercise, and Fitness: social and personal issues*, Washington: Hemisphere Publishing Corp., 1991, Ch. 7.

29 Adapted from Simona Giordano, 'Risks and supervised exercise: the example of anorexia to illustrate a new ethical issue in the traditional debates of medical ethics', *Journal of Medical Ethics* 31, 2005: 15–20.

30 J. Dalgleish, S. Dollery and H. Frankham (eds), *The Health and Fitness Handbook*, London: Longman, 2001, p. 237.

31 D. Pargman, *Understanding Sport Behavior*, Upper Saddle River, NJ: Prentice Hall, 1998, p. 132.

32 D.W. Mickley, 'Medical dangers of anorexia nervosa and bulimia nervosa', in R. Lemberg and L. Cohn (eds), *Eating Disorders: a reference sourcebook*, Phoenix: Oryx Press, 1999, p. 46.

33 Mickley, 'Medical dangers', p. 46.

34 U. Cuntz, G. Frank, P. Lehnert, *et al.*, 'Interrelationships between the size of the pancreas and the weight of patients with eating disorders', *International Journal of Eating Disorders* 27, 2000: 297–303. See also A. Marcos, 'Eating disorders: a situation of malnutrition with peculiar changes in immune system', *European Journal of Clinical Nutrition* 54, 2000 (suppl 1): 61–4S.

35 Katz, 'Eating disorders', pp. 292–312.

36 'Cortical' means that it refers to the outer layer of, in this case, the bone.

37 The trabecula is a transversal partition which divides a cavity.

38 P.S. Mehler and T.D. MacKenzie, 'Treatment of osteopenia and osteoporosis in anorexia nervosa: a systematic review of the literature', *International Journal of Eating Disorders* 42, 2009: 195–201; Suzanne F. Abraham, Bianca Pettigrew, Catherine Boyd and Janice Russell, 'Predictors of functional and exercise amenorrhoea among eating and exercise disordered patients', *Human Reproduction* 21, 1, 2006: 257–61;

P.S. Mehler, A.L. Sabel, T. Watson and A.E. Andersen, 'High risk of osteoporosis in male patients with eating disorders', *International Journal of Eating Disorders* 41, 7, 2008: 666–72; David L. Nichols, Charlotte F. Sanborn and Eve V. Essery, 'Bone density and young athletic women: an update', *Sports Medicine* 37, 11, 2007: 1001–14; K.K. Miller, E.E. Lee, E.A. Lawson, M. Misra, J. Minohan, S.K. Grinspoon, S. Gleysteen, D. Mickley, D. Herzog and A. Klibanski, 'Determinants of skeletal loss and recovery in anorexia nervosa', *Journal of Clinical Endocrinology and Metabolism* 91, 8, 2006: 2931–37; A.D. Divasta, T.J. Beck, M.A. Petit, H.A. Feldman, M.S. LeBoff and C.M. Gordon, 'Bone cross-sectional geometry in adolescents and young women with anorexia nervosa: a hip structural analysis study', *Osteoporosis International* 18, 6, 2007: 797–804; J. Dominguez, L. Goodman, S.S. Gupta, L. Mayer, S.F. Etu, B.T. Walsh, J. Wang, R. Pierson and M.P. Warren, 'Treatment of anorexia nervosa is associated with increases in bone mineral density, and recovery is a biphasic process involving both nutrition and return of menses', *American Journal of Clinical Nutrition* 86, 1, 2007: 92–99; Everett L. Smith and Catherine Gilligan, 'Bone concerns', in Shangold and Mirkin (eds), *Women and Exercise*, Ch. 5, pp. 89–101; A.P. Winston, A.E.F. Alwazeer and M.J.G. Bankart, 'Screening for osteoporosis in anorexia nervosa: prevalence and predictors of reduced bone mineral density', *International Journal of Eating Disorders* 41, 3, 2008: 284–87; Y. Jayasinghe, S.R. Grover and M. Zacharin, 'Current concepts in bone and reproductive health in adolescents with anorexia nervosa', *International Journal of Obstetrics and Gynaecology* 115, 3, 2008: 304–15.
39 Katz, 'Eating disorders', p. 298.
40 Katz, 'Eating disorders', p. 299.
41 Katz, 'Eating disorders'; Mickley, 'Medical dangers', p. 46.
42 Katz, 'Eating disorders', p. 299.
43 N. Kiriike, S. Nishiwaki, T. Nagata, Y. Inoue, K. Inoue and Y. Kawakita, 'Ventricular enlargement in normal weight bulimia', *Acta Psychiatrica Scandinavica* 82, 3, 2007: 264–66; S. Datlof, P.D. Coleman, G.B. Forbes and R.E. Kreipe, 'Ventricular dilation on CAT scans of patients with anorexia nervosa', *American Journal of Psychiatry* 143, 1986: 96–98.
44 Katz, 'Eating disorders', p. 301.
45 WHO, 'Controlling the global obesity epidemic', available at: http://www.who.int/nutrition/topics/obesity/en/ (accessed 25 November 2009).
46 The following account is adapted from Jane Ogden, *The Psychology of Eating: from healthy to disordered behaviour*, Oxford: Blackwell, 2003.
47 Some argue that the Waist–Hip Ratio is a better indicator of health risk. See, for a general idea, http://en.wikipedia.org/wiki/Waist-hip_ratio (accessed 25 November 2009).
48 Ogden, *The Psychology of Eating*, pp. 137 ff.
49 National Institute of Health, *Clinical Guidelines on the Identification, Evaluation, and Treatment of Overweight and Obesity in Adults*, Bethesda, MD: Department of Health and Human Services, National Institutes of Health, National Heart, Lung, and Blood Institute, 1998, pp. 12–20; A. H. Mokdad, Earl Ford, Barbara A. Bowman, *et al.*, 'Prevalence of obesity, diabetes, and obesity-related health risk factors', *Journal of the American Medical Association* 289, 2001: 76–79; A.J. Stunkard and T.A. Wadden (eds), *Obesity: theory and therapy*, 2nd edn, New York: Raven Press, 1993, p. 224.
50 Andrew M. Prentice and Susan A. Jebb, 'Obesity in Britain: gluttony or sloth?', *British Medical Journal* 311, 1995: 437–43.
51 NIDDK, 'Statistics related to overweight and obesity', available at: http://win.niddk.nih.gov/statistics/index.htm (accessed 25 November 2009).
52 Centers for Disease Control and Prevention, 'Prevalence of physical activity, including lifestyle activities among adults, United States, 2000–2001', *Morbidity and Mortality Weekly* 52, 2003: 764–69.

53 For an overview of recommended exercise for health, see the ACSM's 'Physical activity and public health guidelines', available at: http://www.acsm.org/AM/Template.cfm?Section=Home_Page&TEMPLATE=/CM/HTMLDisplay.cfm&CONTENTID=7764 (accessed 25 November 2009).

54 Ogden, *The Psychology of Eating*, p. 147.

55 Ogden, *The Psychology of Eating*, p. 146.

56 Wilfley, Grilo and Brownell, 'Exercise and regulation of body weight', p. 32.

57 Ogden, *The Psychology of Eating*, p. 164.

58 Katz, 'Eating disorders', p. 296.

59 Dana K. Cassell and David H. Gleaves, *The Encyclopedia of Obesity and Eating Disorders*, 3rd edn, New York: Facts on File, 2006, p. 22.

60 Adapted from Howley and Franks, *Fitness Professional's Handbook*, p. 186.

61 For a full account see Katz, 'Eating disorders', pp. 308–9.

62 A fuller account of these can be found in Katz, 'Eating disorders', pp. 308–9.

63 Katz, 'Eating disorders', p. 309; T. Hechler, P. Beumont, P. Marks and S. Touyz, 'How do clinical specialists understand the role of physical activity in eating disorders?', *European Eating Disorders Review* 13, 2, 2005: 125–32.

64 L. Kron, J.L. Katz and H. Weiner, 'The aberrant reproductive endocrinology of anorexia nervosa', in H. Weiner, M.A. Hofer and A.J. Stunkard (eds), *Brain, Behavior and Bodily Disease*, New York: Raven Press, 1981, p. 165. Quoted in Katz, 'Eating disorders'.

65 P.J.V. Beumont, B. Arthur, J.D. Russell and S.W. Touyz, 'Excessive physical activity in dieting disorder patients: Proposals for a supervised exercise program', *International Journal of Eating Disorders* 15, 1994: 21–36; S. Davies, K. Parekh, K. Etelapaa, D. Wood and T. Jaffa, 'The inpatient management of physical activity in young people with anorexia nervosa', *European Eating Disorders Review* 16, 5, 2008: 334–40.

66 Rachel Calogero and Kelly Pedrotty, 'The practice and process of healthy exercise: an investigation of the treatment of exercise abuse in women with eating disorders', *Eating Disorders* 12, 4, 2004: 273–91; Duesund and Skarderud, 'Use the body'.

67 Some authors also advise providing guidelines for instructors, who are also, according to one research, at a higher than normal risk of developing eating disorders. See K. Hoglund and L. Normen, 'A high exercise load is linked to pathological weight control behavior and eating disorders in female fitness instructors', *Scandinavian Journal of Medicine and Science in Sports* 12, 5, 2002: 261–75.

68 Nathan J. Smith, 'Excessive weight loss and food aversion in athletes simulating anorexia nervosa', *Pediatrics* 66, 1 July 1980: 139–42.

69 M.T. Pugliese, F. Lifshitz , G. Grad, P. Fort and M. Marks-Katz, 'Fear of obesity: a cause of short stature and delayed puberty', *The New England Journal of Medicine* 309, 1983: 513–18.

70 William D. McArdle, Frank I. Katch and Victor L. Katch, *Sports and Exercise Nutrition*, Philadelphia: Lippincott Williams and Wilkins, 1999, p. 469.

71 Maria Lucia Bosi and Fatima Palha de Oliveira, 'Bulimic behavior in adolescent athletes', in Jerome S. Rumin (ed), *Eating Disorders and Weight Loss Research*, New York: Nova Science Publishers, 2007, Ch. 3, pp. 67–76, p. 68.

72 Bosi and de Oliveira, 'Bulimic behavior in adolescent athletes', p. 69.

73 Cassell and Gleaves, *The Encyclopedia*, p. 22.

74 See for example, S. Abraham, 'Eating and weight controlling behaviours of young ballet dancers', *Psychopathology* 29, 4, 1996: 218–22; Linda H. Hamilton *et al.*, 'The role of selectivity in the pathogenesis of eating problems in ballet dancers', *Medicine and Science in Sports and Exercise* 20, 6, 1988: 560–65; E.F. Pierce and M.L. Dalent, 'Distortion of body image among elite female dancers', *Perceptual and Motor Skills* 87, 1998: 769–70; C. Rivaldi *et al.*, 'Eating disorders and body image disturbances among ballet dancers, gymnasium users and body builders', *Psychopathology* 36, 5, 2003: 247–54; Diana Schnitt, 'Psychological issues in dancers: an overview', *Journal of Physical Education Recreation and Dance* 61, 9, 1990: 32–34.

75 Cassell and Gleaves, *The Encyclopedia*, p. 55.
76 Ann Rucinski, 'Relationship of body image and dietary intake of competitive ice skaters', *Journal of the American Dietetic Association* 89, 1, 1989: 98–100.
77 Cassell and Gleaves, *The Encyclopedia*, p. 50.
78 Howley and Franks, *Fitness Professional's Handbook*, p. 117; K.A. Beals and N.L. Meyer, 'Female athlete triad update', *Clinics in Sports Medicine* 26, 1, 2007: 69–89.
79 Mona M. Shangold, 'Menstruation and menstrual disorders', in Shangold and Mirkin (eds), *Women and Exercise*, Ch. 9, pp. 152–71, p. 154.
80 Bosi and de Oliveira, 'Bulimic behavior in adolescent athletes', p. 69.
81 Richard F. Ferraro, 'The relationship between types of female athletic participation and female body type', *The Journal of Psychology* 138, 2, 2004: 115–28; see also Gary S. Goldfield, Dan W. Harper and Arthur G. Bluin, 'Are bodybuilders at risk for an eating disorder?', *Eating Disorders* 6, 1998: 133–57; Zali Yager and Jennifer O'Dea, 'The role of teachers and other educators in the prevention of eating disorders and child obesity: what are the issues?', *Eating Disorders* 13, 3, 2005: 261–78.
82 Bosi and de Oliveira, 'Bulimic behavior in adolescent athletes', p. 73.

7 People with eating disorders in the gym

1 For an account of paternalism, see for example T. Beauchamp, 'Paternalism', in W.T. Reich (eds), *Encyclopedia of Bioethics*, revised edn, New York: Simon Schuster and Macmillan, 1995, vol. IV, pp. 1914–20; see also A. Buchanan, 'Medical paternalism', *Philosophy and Public Affairs* 7, Summer 1978: 372.
2 See also Gary S. Goldfield, Dan W. Harper and Arthur G. Blouin, 'Are bodybuilders at risk for an eating disorder?', *Eating Disorders* 6, 1998: 133–57; Lebur Rohman, 'The relationship between anabolic androgenic steroids and muscle dysmorphia: a review', *Eating Disorders* 17, 3, 2009: 187–99; Frederick Grieve, 'A conceptual model of factors contributing to the development of muscle dysmorphia', *Eating Disorders* 15, 1, 2007: 63–80; P.E. Mosley, 'Bigorexia: bodybuilding and muscle dysmorphia', *European Eating Disorders Review* 17, 3, 2009: 191–98; Antonia Baum, 'Eating disorders in the male athlete', *Sports Medicine* 36, 1, 2006: 1–6; H.G. Pope, K.A. Phillips and R. Olivardia, *The Adonis Complex: The secret crisis of male body obsession*, New York: Free Press, 2000.
3 Rosa Behar, 'Gender-related aspects of eating disorders: a psychosocial view', in Jerome S. Rubin (ed) *Eating Disorders and Weight Loss Research*, New York: Nova Science Publishers, 2007, pp. 39–65, p. 39.
4 S. Bem, 'The measurement of psychological androgyny', *Journal of Consulting and Clinical Psychology* 42, 1974: 155–62.
5 Behar, 'Gender-related aspects of eating disorders', p. 42.
6 Behar, 'Gender-related aspects of eating disorders', p. 43. This raises the issue of whether body abuses found in body-building and abuse of drugs like steroids has a common basis with eating disorders. In this book I have not analyzed the relationship between various body abuses in women and men. Systematic research of various types of abuses in the different genders is not available, and it would go beyond the remits of this book. However, on reflection, there are important similarities in the way people might feel dissatisfied with their body, and might attempt to modify it at the expense of their health and safety, and might progress to a stage in which they feel 'compelled' to abuse their body. Whether this abuse takes the form of a diet, of exercise, of cathartic practices, of a search for hypertrophy or even cosmetic surgery, is somehow secondary to the fact that there is a severe dissatisfaction, and that continual control of the body releases that dissatisfaction. It is not the 'sought result' that releases the dissatisfaction, but the continual exercise of control. The analysis of the moral background that might explain various abuses has been provided in

Chapter 5. As I anticipated in the introduction and in other places in this book, what is said of eating disorders and over-exercise can be applied to other forms of body abuse, and the ethical and legal analysis – what we should do and how we should deal with these behaviors in the fitness context – is applicable to other body abuses, in addition to eating disorders.

7　M. Siever, 'Sexual orientation and gender as factors in socioculturally acquired vulnerability to body dissatisfaction and eating disorders', *Journal of Consulting and Clinical Psychology* 62, 2, 1994: 252–60.

8　It is arguable that gender roles discussed in this and other research are not as well defined as they appear to be, and are probably more nuanced than is portrayed. However, the importance of this type of research lies in the fact that it allows us to understand the background of eating-disordered exercisers, and allows us to make sense of seemingly irrational behaviors, which, at a first sight, might appear impossible to understand.

9　Simona Giordano, 'Qu'un Souffle de Vent', *Medical Humanities* 28, 1, 2002: 3–8. See also Simona Giordano, *Understanding Eating Disorders*, Oxford: Oxford University Press, 2005, Chs 5 and 6.

10　For an analysis of feminist studies on eating disorders, see my *Understanding Eating Disorders*. The Bibliography at the end of the volume includes, within the sections Clinical Studies and Philosophical Studies, many bibliographic references that can assist the interested reader to study in more depth the way feminist thinkers have thought of eating disorders and the role of the woman in modern and contemporary society.

11　Ronald S. Manley and Karina M. O'Brien, 'Fitness instructors' recognition of eating disorders and attendant ethical/liability issues', *Eating Disorders* 16, 2008: 103–16. This study, among other things, discusses the fitness instructor's capacity to identify eating-disordered clients, and the ethical and legal issues associated with training an exerciser with eating disorders.

8　Law and professional guidelines

1　Tim Kevan, Dominic Adamson and Stephen Cottrell, *Sports Personal Injury: law and practice*, London: Sweet and Maxwell, 2002, p. 129.

2　Julia Dalgleish, Stuart Dollery and H. Frankham, *The Health and Fitness Handbook*, London: Longman, 2001, p. 326.

3　Kevan, Adamson and Cottrell, *Sports Personal Injury*, p. 9.

4　Mark Lunney and Ken Oliphant, *Tort Law, Text and Material*, Oxford: Oxford University Press, 2008, in particular pp. 111–43.

5　Kevan, Adamson and Cottrell, *Sports Personal Injury*, p. 129.

6　Kevan, Adamson and Cottrell, *Sports Personal Injury*, pp. 130–31.

7　*Rootes v. Skelton* [1968] ALR. 33.

8　Kevan, Adamson and Cottrell, *Sports Personal Injury*, Ch. 3.

9　*Donoghue v Stevenson* [1932] AC 562 at 580; *Agar v Canning* [1965] 54 WWR 302 at 304.

10　*Bolam v Friern Hospital Management Committee* [1957] 1 WLR 582.

11　*Bolam v Friern Hospital Management Committee* [1957] 1 WLR 582 at 587–88. Commentary can be found in Margaret Brazier and Emma Cave, *Medicine Patients and the Law*, London: Butterworths, 2007, p. 161.

12　*Phillips v William Whiteley Ltd* [1938] 1 All ER 566. See Brazier and Cave, *Medicine Patients and the Law*, p. 159.

13　*Bolam v Friern Hospital Management Committee* [1957] 1 WLR 582 at 587–88.

14　*Bolitho v Hackney Health Authority* [1997] 3 WLR 1151.

15　Brazier and Cave, *Medicine Patients and the Law*, p. 162.

16   [1998] AC 232 at 242, HL.
17   [1998] AC 232 at 243, HL.
18   Brazier and Cave, *Medicine Patients and the Law*, p. 164.
19   *Barnett v Chelsea & Kensington Hospital Management Committee* [1968] 1 All ER 1068.
20   Adapted from Tony Hope, Julian Savulescu and Judith Hendrick, *Medical Ethics and Law: the core curriculum*, 2nd edn, London: Churchill Livingstone, 2008, p. 53.
21   The Occupiers Liability Act 1957 can be found at http://www.statutelaw.gov.uk/ (accessed 25 November 2009).
22   The issue of proper screening is a difficult topic. We see at the end of the chapter, that the ACSM proposes several ways of screening clients, and these are recommended. However, in normal studio situations, studio instructors do not see PAR-Qs or other screening forms. This is because in some gyms it is expected that the managers do this before the client is given membership and access to the classes. In other gyms classes are pay-as-you-go, and the instructor does not have the time to screen each new participant before the beginning of the session. Finally, there is a broader issue of public health benefits versus costs. It is probably overall good for the population to have easily accessible fitness facilities, places where we easily get our training, without doing health and lifestyle screenings every time. Most swimming pools are open to the general public in the UK and in other countries. The overall benefit in terms of public health probably outweighs the hazards of this system. A system where every supervised fitness activity must be preceded by a thorough health and lifestyle screening would probably reduce to some extent people's access to fitness.
23   Unreported December 15 1998 (CA (Qld)). Quoted in Phillips Fox, 'Sports and Gaming law: Australia: power walker gets marching orders', *International Legal Practitioner*, 24, 3, 1999: 71–72.
24   Kevan, Adamson and Cottrell, *Sports Personal Injury*, p. 131.
25   Kevan, Adamson and Cottrell, *Sports Personal Injury*, p. 132.
26   Alastair Mullis and Ken Oliphant, *Torts*, London: Macmillan, 2003, p. 90. Lunney and Oliphant, *Tort Law*, pp. 111–43.
27   Kevan, Adamson and Cottrell, *Sports Personal Injury*, p. 10.
28   It should be noted that a simple notice excluding liability (for example, I put a notice in my studio which says I shall not be liable for any personal injury suffered by participants) cannot exclude liability for personal injury. I owe this observation to Margaret Brazier.
29   Kevan, Adamson and Cottrell, *Sports Personal Injury*, p. 10.
30   Feeding in cases of anorexia can be enforced by law in England and Wales under the Mental Health Act 1983 S.3, regardless of determination of capacity. However, in a number of cases brought before the Courts, it was also argued that anorexia creates a compulsion to refuse meaningful treatment and food, and that anorexia nervosa jeopardizes people's competence. *Re W* (a minor) (medical treatment court's jurisdiction) [1993] Fam 64 at 81, [1992] 4 All ER 627 at 637; *Re C (a minor) (detention for medical treatment)* [1997] 2 FLR 180 (Fam Div). A commentary can be found in Ian Kennedy and Andrew Grubb, *Medical Law*, London, Edinburgh, Dublin: Butterworths, 2000, pp. 639–41.
31   I have elsewhere argued that the arguments brought by law to justify compulsory treatment are not substantiated and are conceptually flawed (see Chapter 9), but for the sake of the argument we need to assume that this is what could be argued in case of litigation.
32   Simona Giordano, *Understanding Eating Disorders*, Oxford: Oxford University Press, 2005.
33   [1990] 2 All ER 346.
34   See Duncan Fairgrieve, *State Liability in Tort*, Oxford: Oxford University Press, 2003, pp. 181–82, and footnote 157.

35 [1999] 3 WLR 363.

36 Vivienne Harpwood, *Negligence in Healthcare, Clinical Claims and Risk in Context*, London: Informa Publishing Group, 2001, p. 157.

37 Law Reform (Contributory Negligence) Act 1945. See Harpwood, *Negligence in Healthcare*, p. 158.

38 Brazier and Cave, *Medicine Patients and the Law*, p. 168.

39 Available at: www.exerciseregister.org (accessed 25 November 2009).

40 The previous version has been removed from the REPs website. The interested reader can contact the Register of Exercise Professionals at: www.exerciseregister.org/REPsContact.html (accessed 25 November 2009).

41 In some instances, it seems that the right to self-determination can be limited in the best interests of the client. Fitness professionals are for example advised to keep a 'paternalistic' attitude in some cases. It is for example suggested that well-cushioned supportive shoes be *compulsory*; or that women during pregnancy provide medical permission to the fitness instructor prior to exercise. G. Egger, N. Champion and A. Bolton, *The Fitness Leader's Handbook*, London: A & C Black, 2002, pp. 116, 168.

42 Available at: http://www.acefitness.org/getcertified/certified-code.aspx (accessed 25 November 2009).

43 Brazier and Cave, *Medicine Patients and the Law*, Ch. 4.

44 British Medical Association, *Medical Ethics Today: the BMA's handbook of ethics and law*, London: BMJ, 2004, p. 181.

45 Where GPs or other medical professionals liaise with some sports and fitness centers, and with instructors (sometimes called 'qualified GP referral instructors') and prescribe some types of exercise to their clients. GPs in these cases refer their patients to the fitness professionals.

46 American College of Sports Medicine, *ACSM's Guidelines for Exercise Testing and Prescription*, 7th edn, Philadelphia: Lippincott Williams and Wilkins, 2006; American College of Sports Medicine, *ACSM's Resource Manual for Guidelines for Exercise Testing and Prescription*, 5th edn, Philadelphia: Lippincott Williams and Wilkins, 2006; American College of Sports Medicine, *ACSM's Health/Fitness Facility Standards and Guidelines*, 3rd edn, Champaign, IL: Lippincott Williams and Wilkins, 2007; American College of Sports Medicine and American Heart Association, 'ASCM/AHA joint position statement: Recommendations for cardiovascular screening, staffing and emergency policies at health/fitness facilities', *Medicine and Science in Sports and Exercise* 30, 6, 1998:1009–18.

47 Morc Coulson, *The Fitness Instructor's Handbook: a complete guide to health and fitness*, London: A&C Black, 2007, p. 168.

48 Adapted from Coulson, *Fitness Instructor's Handbook*, p. 170.

49 A fuller account of health appraisal and discussion of ACSM and ACC guidelines for testing and exercise prescription can be found in Edward T. Howley and B. Don Franks, *Fitness Professional's Handbook*, 5th edn, Champaign, IL: Human Kinetics, 2007, pp. 21–38.

50 Volume (V) per time, oxygen (O_2), maximum (Max). It is calculated at maximal effort, and is expressed in liters of oxygen per minute or in milliliters of oxygen per kilogram of body weight. It is the maximal capacity of a person to use oxygen during exercise. A standard measure, taken to be a rough indication of VO2Max, is 220 – [age].

51 R. Gibbons, G Balady, J. Bricker, B. Chaitman, G. Fletcher, V. Froelicher, F. Mark *et al.*, 'ACC/AHA 2002 guideline update to exercise testing: Summary article. A report of the American College of Cardiology/American Heart Association task force on practice guidelines (Committee to update the 1997 exercise testing guidelines)', *Journal of the American College of Cardiology* 40, 8, 2002: 1531–40. Available at: http://circ.ahajournals.org/cgi/content/full/106/14/1883 (accessed 25 November 2009).

52 Gibbons, Balady, Beasely *et al.*, 'ACC/AHA 2002 guideline'.

53 Dalgleish, Dollery and Frankham, *Health and Fitness Handbook*, p. 141.
54 Walter Vandereycken, 'Denial of illness in anorexia nervosa – a conceptual review Part 1. Diagnostic significance and assessment', *European Eating Disorders Review* 14, 2006: 341–53; Walter Vandereycken, 'Denial of illness in anorexia nervosa – a conceptual review. Part 2. Different forms and meanings', *European Eating Disorders Review* 14, 2005: 352–68; Walter Vandereycken, 'Denial and concealment of eating disorders: a retrospective study', *European Eating Disorders Review* 16, 2008: 109–14.
55 Available at: http://www.fitnessstandards.org/Practitioners/ethics.html (accessed 25 November 2009).
56 Available at: http://www.ncsf.org/boardcert/ethics.aspx (accessed 25 November 2009).

9 Ethical issues

1 'Down the rabbit hole', by Stephanie Mayer, courtesy of Caring Online. Available at: http://www.caringonline.com/feelings/byvictims/mayer.htm (accessed 25 November 2009).
2 For references see Simona Giordano, *Understanding Eating Disorders*, Oxford: Oxford University Press, Oxford, 2005, in particular Chs 1, 2 and 11 and 13. The Bibliography at the end of the volume also contains numerous references.
3 'Celebrities', available at: http://caringonline.com/eatdis/people.html (accessed 25 November 2009).
4 Walter Vandereycken, 'Denial and concealment of eating disorders: a retrospective study', *European Eating Disorders Review*, 16, 2008: 109–14.
5 Wilfried Ver Eecke, *Denial, negation, and the forces of the negative: Freud, Hegel, Lacan, Spitz, and Sophocles*, New York: Suny, 2006, p. 130.
6 There is a distinction to draw here between state paternalism and paternalism in individual transactions. State paternalism is more difficult to justify in these cases. *Restrictions of freedom of action* in individual transactions, instead, are more common. For example: it is common practice to *oblige* gym users to have an 'induction'. This is like saying: you can only use the gym equipment if you accept that I train you so that you know how to use it. And I accept you in the gym only provided that you comply with various conditions. I again owe this observation to John Coggon. The suggestion of this chapter is that these *limitations of freedom* are not *forms of paternalism*, but could more appropriately be subsumed under *contractualism*. Paternalism entails limitations of freedom primarily aimed at protecting or promoting the person's welfare (see also Chapter 7). Restrictions of freedom of action in individual transactions are not necessarily aimed, at least primarily, at protecting and promoting the person's welfare. Mandatory induction, in this example, can be understood in terms of the contract, as I argue in this chapter, between clients and professionals, of legal responsibility on the part of the fitness professionals and gym owners, and not in terms of 'doing what is good for the client'. The difference between paternalism and contractualism will become clear in the course of the chapter.
7 T. Beauchamp, 'Paternalism', in W.T. Reich (ed), *Encyclopedia of Bioethics*, revised edn, New York: Simon Schuster and Macmillan, 1995, pp. 1914–20, G. Dworkin, 'Paternalism', *Monist* 56, Jan. 1972: 64–84, J. Rawls, *A Theory of Justice*, Oxford: Oxford University Press, 1972, p. 249.
8 J. Feinberg, 'Legal paternalism', *Canadian Journal of Philosophy*, 1, 1971: 105–24. For a fuller account of the various theories of paternalism, and how they can be applied to the psychiatric field, see Giordano, *Understanding Eating Disorders*, Ch. 2.
9 *Sidaway v Board of Governors of the Bethlem Royal Hospital and the Maudsley Hospital* [1985] 1 All ER 643 at 509 b per Lord Templeman; see also *R v Blame* [1975] 3 All ER 446.

10 Aurora Plomer, *The Law and Ethics of Medical Research: international bioethics and human rights*, London: Cavendish Publishing, 2005, p. 58.

11 J. S. Mill, *On Liberty*, New York: Cambridge University Press, 1989, p. 13.

12 One could even argue that, in these cases, there is a *prima facie* moral obligation to intervene (not just a moral right, as I argued here). *Prima facie* means 'to be accepted until there is proof otherwise'. For example, where the burdens for me are far too high, arguably this moral obligation ceases. This would involve an argument as to which risks and burdens are too high, and which are low enough for this moral obligation to continue to exist. This all goes beyond the remits of this book.

13 Mill, *On Liberty*, pp. 106–7.

14 For a full account of these two forms of paternalism, and on the philosophical debates on the concepts of autonomy and competence, see Giordano, *Understanding Eating Disorders*, Ch. 2.

15 Mill, *On Liberty*, p. 13.

16 I. Berlin, 'Two concepts of liberty', in I. Berlin, *Four Essays on Liberty*, Oxford: Oxford University Press, 1969, p. 131.

17 I use the terms autonomy, competence and capacity, or decision-making capacity, as equivalent. Autonomy is a rather philosophical term, whereas capacity/competence is used more often in clinical and legal settings. Some see autonomy as a feature of persons, and competence as a feature of actions and choices. Some distinguish between competence and decision-making capacity. These subtle distinctions are not relevant to our discourse and the terms are not differentiated here. All these terms, here, refer to the capacity of the individual to make 'their own' decisions without undue influences (such as lack of information, lack of understanding of relevant facts, psychotic crises or the effect of psychoactive drugs).

18 *Re W (a minor) (medical treatment court's jurisdiction)* [1993] Fam 64 at 81, [1992] 4 All ER 627 at 637. J.K. Mason, R.A. McCall Smith and G.T. Laurie, *Law and Medical Ethics*, 6th edn, London: Butterworths, 2002, p. 633.

19 *Re C (a minor) (detention for medical treatment)* [1997] 2 FLR 180 (Fam Div). Ian Kennedy and Andrew Grubb, *Medical Law*, London, Edinburgh, Dublin: Butterworths, 2000, pp. 639–41.

20 A fuller account can be found in Simona Giordano, 'In defence of autonomy in psychiatric healthcare', *Turkish Journal of Medical Ethics* 9, 2, 2001: 59–66; Giordano, *Understanding Eating Disorders*, Ch. 3.

21 Or sometimes a person.

22 I have also pointed this out in my book *Understanding Eating Disorders*.

23 I also owe this to Harry Lesser.

24 I have analyzed the good potential of psychiatric diagnosis in my *Understanding Eating Disorders*, Chs 3 and 11.

25 It should also be noted that in England and Wales, the Mental Health Act 1983 (and 2007) establishes that people hospitalized coercively for reasons relating to their mental illness shall receive treatment for their mental illness regardless of whether they give, or are capable of giving, their consent to treatment. Their competence to consent to, or refuse, psychiatric treatment, is irrelevant in these cases (S. 63). In the case of anorexia, it has been discussed in the courts whether food and nutrition could be treatment for anorexia, as opposed to treatment for malnutrition, and thus, whether this treatment could be lawfully imposed under the Mental Health Act. It was established that it is. For full account and critique, see my *Understanding Eating Disorders*, chapter 11 in particular.

26 And unnecessary.

27 Simona Giordano, 'Choosing death in cases of anorexia nervosa: should we ever let people die from anorexia nervosa?', in Charles Tandy (ed), *Death and Anti-death*, Volume 5, Palo Alto, CA: Ria University Press, 2008, Ch. 9.

28 Sir Karl Popper, *The Open Society and Its Enemies*, first published in London: Routledge, 1945, Ch. 5.

29 J.J.C. Smart and B. Williams, *Utilitarianism: for and against*, Cambridge: Cambridge University Press, 1996, p. 28.

30 Smart and Williams, *Utilitarianism*, p. 28.

31 I am not discussing the relationship between Beneficence and Non-Maleficence as this is not strictly relevant to our discourse.

32 Jack L. Katz, 'Eating disorders', in Mona M. Shangold and Gabe Mirkin (eds), *Women and Exercise, Physiology and Sports Medicine*, Philadelphia: Davis Company, 1994, pp. 292–312.

33 G. Egger, N. Champion and A. Bolton, *The Fitness Leader's Handbook*, London: A&C Black, 2002, pp. 14, 18.

34 Egger, Champion and Bolton, *Fitness Leader's Handbook*, pp. 16, 116.

35 M. Selvini Palazzoli, S. Cirillo, M. Selvini and A.M. Sorrentino, *Ragazze anoressiche e bulimiche, la terapia familiare*, Milan: Cortina, 1998, pp. 96–97.

36 Günter Rathner, 'A plea against compulsory treatment of anorexia nervosa patients', in Pierre J.V. Beumont and Walter Vandereycken (eds), *Treating Eating Disorders: ethical, legal and personal issues*, New York: New York University Press, 1998, p. 198.

37 Rathner, 'A plea against compulsory treatment', p. 198.

38 Walter Vandereycken and Maarten Vanstecnkiste, 'Let eating disorder patient decide: providing choice may reduce early drop-out from inpatient treatment', *European Eating Disorders Review* 17, 2009: 177–83.

39 Margery Gans and Willam B. Gunn, 'End stage anorexia: criteria for competence to refuse treatment', *International Journal of Law and Psychiatry* 26, 6, 2003: 677–95.

40 Jon Mitchell and Robyn Eisenbach, 'The recreational therapist role in prescribing exercise to the eating disorder patient', available at: http://www.recreationtherapy.com/articles/mitchell.htm (accessed 25 November 2009).

41 J. Dalgleish, S. Dollery and H. Frankham (eds), *The Health and Fitness Handbook*, London: Longman, 2001, p. 332.

42 Definition from: http://en.wikipedia.org/wiki/Customer_service (accessed 25 November 2009).

43 The notions of health and welfare are of course open to interpretation. Yet, breaking limbs and risking a cardiovascular collapse cannot be included in the notion, by any reasonable account. The eating disorders sufferer has often a strong interest in taking exercise, and exercise can be beneficial in many ways to them (psychologically and physically). For this reason, exercise should not be demonized (see Chapter 10), but utilized in a competent manner by qualified professionals.

44 Simona Giordano, 'Gender atypical organisation in children and adolescents: ethico-legal issues and a proposal for new guidelines', *International Journal of Children's Rights* 15, 3–4, 2007: 365–90.

45 Simona Giordano, 'Lives in a chiaroscuro. Should we suspend the puberty of children with Gender Identity Disorder?', *Journal of Medical Ethics* 34, 8, 2008: 580–86.

46 Airedale NHS Trust v Bland [1993] 1 All ER 821 (HL); John C. Hall, 'Acts and omissions', *The Philosophical Quarterly* 39, 157, 1989: 399–408; N.D. Husak, 'Omission, causation and liability', *The Philosophical Quarterly* 30, 121, 1980: 318–26.

47 I have argued previously that eating disorders are better regarded as existential conditions rather than 'psychiatric illnesses'. Some could object here that, by suggesting that fitness professionals advise clients to contact GPs or other clinical services, in contradiction to my previous argument, I advise treating eating disorders as 'illnesses'. A GP (a medic), in fact, is likely to refer the patient to a psychiatrist (a medic) or a clinical psychologist. So, the objection might further go, if these medics are likely to work to a medical model, is it not irresponsible for a fitness professional to suggest mental health professionals who (a) might see the person as having an illness, and (b) might treat her as such? I owe these observations to Adam James. The

answer to these objections is that, although eating disorders are an existential condition, they are also a complex condition, and the physical perils associated with abnormal eating are serious and potentially life-threatening. There is thus a medical side to eating disorders, which cannot and should not be ignored in the sports and fitness context. Fitness professionals who are not qualified to supervise these clients would thus be irresponsible to carry out tasks that go beyond their competencies. Fitness professionals would rightly get advice from someone who is more knowledgeable. They would rightly advise the client to work together with medics, who can control the physical sequel associated with abnormal eating, and would rightly take their advice, but cannot be held morally responsible for what GPs or other healthcare professionals, rightly involved in the management of the sufferer, decide to do.

48 Morc Coulson, *The Fitness Instructor's Handbook: a complete guide to health and fitness*, London: A&C Black, 2007, p. 167.

49 D. W. Mickley, 'Medical dangers of anorexia nervosa and bulimia nervosa', in R. Lemberg and L. Cohn (eds), *Eating Disorders: a reference sourcebook*, Phoenix: Oryx Press, 1999, p. 47.

50 I owe this observation to Iain Brassington.

51 NSCA Professional Standards and Guidelines Task Force, *Strength and Conditioning: professional standards & guidelines*, Colorado Springs, CO: NSCA, 2001, p. 38. Available at: http://www.nsca-lift.org/Publications/SCStandards.pdf (accessed 25 November 2009).

10 Recommendations and conclusions

1 Mona M. Shangold, 'Beyond the exercise prescription: making exercise a way of life', *The Physician and Sports Medicine* 26, 11, November 1998. Available at: http://www.physsportsmed.com/index.php?art=psm_11_1998?article=1187 (accessed 25 November 2009).

2 K. Höglund and L. Normen, 'A high exercise load is linked to pathological weight control behaviour and eating disorders in female fitness instructors', *Scandinavian Journal of Medicine and Science in Sports* 12, 5, 2002: 261–75.

3 This is a general legal principle. The US and the UK have different understanding of the circumstances in which, once someone has breached the duty of care, the person who suffered harm can be compensated.

4 In an interesting study on the role of teachers, it is pointed out that exercise instructors who suffer body dissatisfaction and eating/exercise disorders might intentionally or unintentionally convey inappropriate eating and exercise attitudes to their school students. Zali Yager and Jennifer O'Dea, 'The role of teachers and other educators in the prevention of eating disorders and child obesity: what are the issues?', *Eating Disorders* 13, 3, 2005: 261–78.

5 See, however, the newspaper article "Gym let me join when I was anorexic . . . even though one step on a treadmill could have killed me", *Daily Mail*, 29 July 2009. In this case, Ms. Jessica Bennington claims that a well-known fitness center in the UK allowed her to join the gym, thus endangering her health, although she was clearly anorexic. It is not clear whether she will sue the fitness center and exactly on what grounds. I wish to thank Sheelagh McGuinness for pointing this article out to me.

Index